AF572720

Developments in Petroleum Science, 29

compressibility of sandstones

DEVELOPMENTS IN PETROLEUM SCIENCE
Advisory Editor: G.V. Chilingarian

Volumes 1, 3, 4, 7 and 13 are out of print.

2. W.H. FERTL
ABNORMAL FORMATION PRESSURES
5. T.F. YEN and G.V. CHILINGARIAN (Editors)
OIL SHALE
6. D.W. PEACEMAN
FUNDAMENTALS OF NUMERICAL RESERVOIR SIMULATION
8. L.P. DAKE
FUNDAMENTALS OF RESERVOIR ENGINEERING
9. K. MAGARA
COMPACTION AND FLUID MIGRATION
10. M.T. SILVIA and E.A. ROBINSON
DECONVOLUTION OF GEOPHYSICAL TIME SERIES IN THE EXPLORATION FOR OIL AND NATURAL GAS
11. G.V. CHILINGARIAN and P. VORABUTR
DRILLING AND DRILLING FLUIDS
12. T.D. VAN GOLF-RACHT
FUNDAMENTALS OF FRACTURED RESERVOIR ENGINEERING
14. G. MÓZES (Editor)
PARAFFIN PRODUCTS
15A O. SERRA
FUNDAMENTALS OF WELL-LOG INTERPRETATION
1. THE ACQUISITION OF LOGGING DATA
15B O. SERRA
FUNDAMENTALS OF WELL-LOG INTERPRETATION
2. THE INTERPRETATION OF LOGGING DATA
16. R.E. CHAPMAN
PETROLEUM GEOLOGY
17A E.C. DONALDSON, G.V. CHILINGARIAN and T.F. YEN
ENHANCED OIL RECOVERY, I
FUNDAMENTALS AND ANALYSES
18A A.P. SZILAS
PRODUCTION AND TRANSPORT OF OIL AND GAS
A. FLOW MECHANICS AND PRODUCTION
second completely revised edition
18B A.P. SZILAS
PRODUCTION AND TRANSPORT OF OIL AND GAS
B. GATHERING AND TRANSPORTATION
second completely revised edition
19A G.V. CHILINGARIAN, J.O. ROBERTSON Jr. and S. KUMAR
SURFACE OPERATIONS IN PETROLEUM PRODUCTION, I
19B G.V. CHILINGARIAN, J.O. ROBERTSON Jr. and S. KUMAR
SURFACE OPERATIONS IN PETROLEUM PRODUCTION, II
20. A.J. DIKKERS
GEOLOGY IN PETROLEUM PRODUCTION
21. W.F. RAMIREZ
APPLICATION OF OPTIMAL CONTROL THEORY TO ENHANCED OIL RECOVERY
22. E.C. DONALDSON, G.V. CHILINGARIAN and T.F. YEN (Editors)
MICROBIAL ENHANCED OIL RECOVERY
23. J. HAGOORT
FUNDAMENTALS OF GAS RESERVOIR ENGINEERING
24. W. LITTMANN
POLYMER FLOODING
25. N.K. BAIBAKOV and A.R. GARUSHEV
THERMAL METHODS OF PETROLEUM PRODUCTION
26. D. MADER
HYDRAULIC PROPPANT FRACTURING AND GRAVEL PACKING
27. G. DA PRAT
WELL TEST ANALYSIS FOR NATURALLY FRACTURED RESERVOIRS
28. E.B. NELSON (Editor)
WELL CEMENTING

Developments in Petroleum Science, 29

compressibility of sandstones

ROBERT W. ZIMMERMAN

Earth Sciences Division, Lawrence Berkeley Laboratory, University of California, Berkeley, CA 94720, U.S.A.

ELSEVIER — Amsterdam — Oxford — New York — Tokyo 1991

ELSEVIER SCIENCE PUBLISHERS B.V.
Sara Burgerhartstraat 25
P.O. Box 211, 1000 AE Amsterdam, The Netherlands

Distributors for the United States and Canada:

ELSEVIER SCIENCE PUBLISHING COMPANY INC.
655, Avenue of the Americas
New York, NY 10010, U.S.A.

ISBN 0-444-88325-8

This book is printed on acid-free paper.

Printed in The Netherlands

Preface

This book presents a comprehensive treatment of the elastic volumetric response of sandstones to variations in stress. The theory and data presented apply to the deformations that occur, for example, due to withdrawal of fluid from a reservoir, or due to the redistribution of stresses caused by the drilling of a borehole. Deformations that occur over geologic time scales are in general excluded, since such processes involve chemical and thermal effects in addition to purely mechanical deformation. Note that while the data presented to illustrate the principles discussed in the book are exclusively for consolidated sandstones, most of the theoretical relationships, bounds, etc. that are discussed are applicable to all rock-fluid systems.

Part One establishes a general framework for the study of the volumetric response of sandstones to applied stresses. Chapter 1 discusses the various compressibility coefficients that can be defined for porous rocks. Relationships between these compressibilities are discussed in Chapter 2, and bounds on the numerical values of the compressibilities are discussed in Chapter 3. Chapter 4 treats the effective stress coefficients, which describe the relative effects of pore pressure and confining pressure. Chapter 5 covers the integrated stress-strain relationships for sandstones, in a fully nonlinear context. Undrained compression and induced pore pressures are treated in Chapter 6. Chapter 7 gives a brief introduction to the theory of poroelasticity, in effect extending the discussion in the previous chapters to situations involving deviatoric stresses.

Part Two presents, for the first time in book form, a detailed discussion of the relationship between compressibility and pore structure. Chapter 8 discusses mathematical models of tubular pores, Chapter 9 treats "crack-like" models of imperfectly bonded grain contacts, and Chapter 10 covers the modeling of pores as spheroids. Chapter 11 discusses the effect of pore shape on the overall effective bulk and shear moduli. Methods for relating the stress-strain behavior of a sandstone to the distribution of the aspect ratios of its pores are described in Chapter 12. Part Three (Chapter 13) presents brief discussions of experimental methods used to measure compressibility in the laboratory.

Both SI and British engineering units are used in this book; in most cases, data is given in both systems of measurement. Stresses and elastic moduli, each of which have dimensions of "pressure", are usually expressed in Megapascals (MPa) or pounds per square inch (psi). The various compressibilities have dimensions of 1/pressure, and so are expressed in units of 1/MPa or 1/psi; the two systems of measurement are related through the equation 1 MPa = 145 psi.

I gratefully express my thanks to Wilbur Somerton, my Ph.D. thesis advisor, and to Michael King and Neville Cook, members of my thesis committee, for all that I learned from them about rock mechanics. I also thank George Chilingarian, the editor of this series, and Jacques Kiebert of Elsevier, for their work in helping this book to be published. Finally, I thank Melanie, without whose encouragement and support this book would not have been written.

CONTENTS

PART ONE: COMPRESSIBILITY AND STRESS

Chapter 1. Definitions of Porous Rock Compressibilities

Compressibility is the parameter that quantifies the relationship between the pressure exerted on a body and the resulting change in its volume. A non-porous material has a single compressibility, C, defined by

$$C = \frac{-1}{V^i}\frac{dV}{dP} , \tag{1.1}$$

where V is the volume of the body, P is the hydrostatic pressure exerted over its outer surface, and the superscript i indicates the initial, stress-free value. In definition (1.1), as well as all of those which follow, it is implicitly assumed that the temperature is held constant as the pressure is being varied. Other common notations used for compressibility are β and κ. The pressure in eqn. (1.1) is considered to be positive if it is compressive (Fig. 1.1), which is the usual thermodynamic convention. Since a compressive stress will reduce the volume, the minus sign in definition (1.1) is used so that the compressibility will be a positive number. The volumetric strain can be defined as $d\varepsilon = dV/V^i$, so that decreases in volume correspond to negative strains. Eqn. (1.1) can then be written in the form $d\varepsilon = -CdP$.

In contrast to the single pressure-volume relationship that exists exists for a non-porous material, the situation is more complicated for a porous rock, since it can be subjected to an external "confining" pressure, as well as an internal "pore" pressure that acts over the surfaces of its pore walls (Fig 1.2). It is also necessary to account for at least two different volumes, the bulk volume and the pore volume. As shown in Fig. 1.2, the bulk volume V_b is defined as the volume that would be measured if the presence of the pores were ignored. The pore volume V_p is defined as that part of the bulk volume which is not occupied by rock minerals. These two volumes are related by

$$V_b - V_p = V_r , \tag{1.2}$$

where V_r is the volume occupied by the mineral grains. The relative amounts of pore space and minerals in the rock are usually quantified by geophysicists and petroleum engineers by the

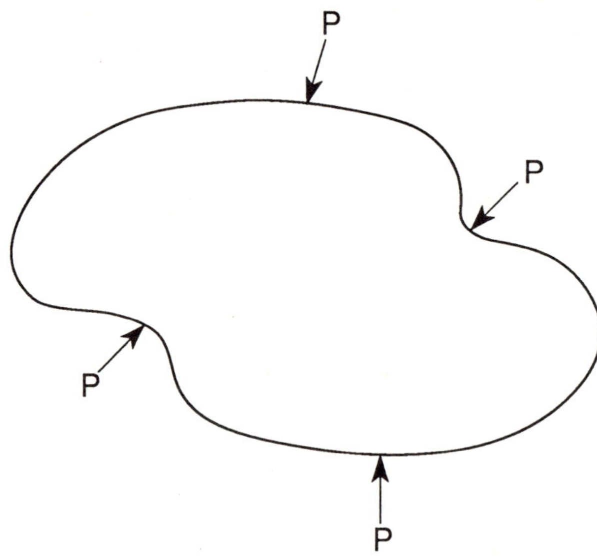

Fig. 1.1. Compressive stress of magnitude P acting over the outer surface of a non-porous body of volume V. Compressive stresses are considered positive, and lead to a decrease in the volume.

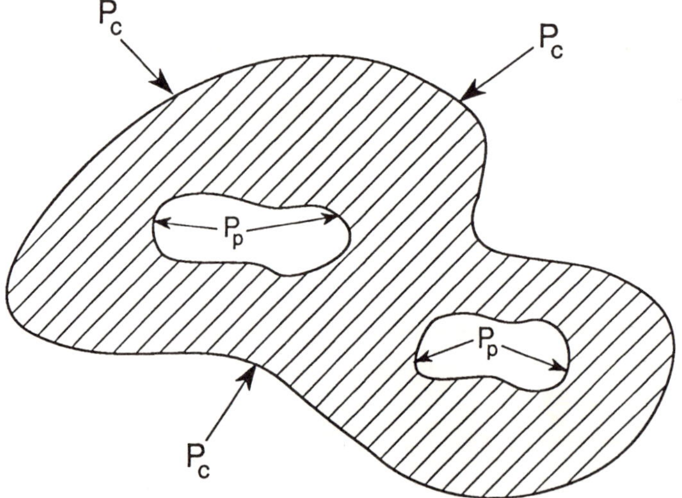

Fig. 1.2. Confining pressure P_c and pore pressure P_p acting on a porous body.

porosity ϕ, defined by

$$\phi = V_p / V_b \ . \tag{1.3}$$

Civil engineers often use the void ratio e, defined by

$$e = V_p / V_r = \frac{V_p}{V_b - V_p} = \frac{\phi}{1-\phi} \ . \tag{1.4}$$

Note that while the porosity is restricted, by definition, to the range $0 \le \phi < 1$, the void ratio can take on any positive value. Although the void ratios of some uncompacted near-surface soils are greater than unity, since their porosities may exceed 50%, values of e for consolidated sandstones will typically be less than one, since their porosities will be less than 50%.

Since there are two independent volumes and two pressures that can be varied, four different compressibilities can be associated with a porous rock. Each of these porous rock compressibilities relates changes in either the pore volume V_p or the bulk volume V_b to changes in the pore pressure P_p or the confining pressure P_c. Using a notation in which the first subscript indicates the relevant volume change, and the second subscript indicates the pressure which is varied, these compressibilities can be defined as follows [Zimmerman *et al.*, 1986a]:

$$C_{bc} = \frac{-1}{V_b^i} \left(\frac{\partial V_b}{\partial P_c} \right)_{P_p} , \tag{1.5}$$

$$C_{bp} = \frac{1}{V_b^i} \left(\frac{\partial V_b}{\partial P_p} \right)_{P_c} , \tag{1.6}$$

$$C_{pc} = \frac{-1}{V_p^i} \left(\frac{\partial V_p}{\partial P_c} \right)_{P_p} , \tag{1.7}$$

$$C_{pp} = \frac{1}{V_p^i} \left(\frac{\partial V_p}{\partial P_p} \right)_{P_c} . \tag{1.8}$$

The derivative in the definition of C_{bc}, for example, is to be interpreted as the partial derivative of the bulk volume with respect to the confining pressure, with the pore pressure (and temperature) held constant. The pore and confining pressures are here assumed to be mathematically independent variables; physically, this corresponds to so-called "drained compression", in which variations in confining pressure do not cause the pore pressure to change. Under some conditions of loading, this will not be true, although the definitions given in eqns. (1.5-1.8) will still be useful for deriving an expression for the "undrained compressibility" (see Chapter 6). The bulk and pore strain increments can be expressed in terms of the porous rock compressibilities as follows:

$$d\varepsilon_b = -C_{bc}\, dP_c + C_{bp}\, dP_p \ , \tag{1.9}$$

$$d\varepsilon_p = -C_{pc}\, dP_c + C_{pp}\, dP_p \ . \tag{1.10}$$

The two compressibilities defined by eqns. (1.5) and (1.6), namely C_{bc} and C_{bp}, are referred to as bulk compressibilities, since they involve changes in the bulk volume of the rock. The bulk compressibility C_{bc} is analogous to the compressibility of a non-porous material as defined by eqn. (1.1). This compressibility finds use in large-scale tectonic calculations, as well as in wave-propagation problems. Insofar as a fluid-saturated sandstone can be treated as a homogeneous material, C_{bc} will be its effective bulk compressibility. The other bulk compressibility, C_{bp}, which was called the "pseudo-bulk compressibility" by Fatt [1958a], reflects the influence of pore pressure on the bulk volume. This compressibility is useful in subsidence calculations, for instance [Geertsma, 1973]. As fluid is withdrawn from a petroleum or groundwater reservoir, the pore pressure will decline, and the total volume occupied by the reservoir will decrease, in accordance with eqn. (1.9). The consequent subsidence of the ground surface above the reservoir can be on the order of tens of meters, and has been known to cause extensive damage to buildings and other surface structures. Well-known examples of oilfield subsidence include the Wilmington-Long Beach field in Southern California [Gilluly and Grant, 1949], and the Ekofisk field in the North Sea [Johnson *et al.*, 1989].

The other two compressibilities, C_{pc} and C_{pp}, are pore compressibilities, and express the effect of pressure variations on the volume of void space contained in the rock. Hall [1953] referred to C_{pc} as the "formation compaction" coefficient, and to C_{pp} as the "effective pore compressibility". The pore compressibility C_{pp} is used in reservoir analysis, since it reflects the volume of excess pore fluid that can be stored in the pore space due to an increase in the pore pressure. This compressibility is added to that of the reservoir fluid, C_f, in order to represent the "reservoir compressibility" term that is used in the basic equation of reservoir analysis. For single-phase liquid flow, the governing equation of pressure transient analysis for reservoirs is [Matthews and Russell, 1967]

$$\nabla^2 P_p = \frac{\phi \mu C_t}{k} \frac{\partial P_p}{\partial t} , \tag{1.11}$$

where ∇^2 is the Laplacian operator, μ is the viscosity of the pore fluid, k is the formation permeability, and $C_t = C_f + C_{pp}$ is the total "compressibility" of the rock/fluid system.

The pore compressibility C_{pc} has relevance to laboratory determinations of the *in situ* pore volume of a reservoir rock. When a rock core is brought to the surface, the confining stress that had been acting on it due to the overburden is relaxed. In order to correctly determine the *in situ* pore volume, the core is placed in a pressure vessel, and the confining stress is increased to its *in situ* value, P_c. Such experiments therefore involve the pore compressibility C_{pc}, which relates the change in pore volume to the confining pressure.

Since positive confining pressures will decrease the volumes V_p and V_b, while positive pore pressures will increase V_p and V_b, the definitions of the two compressibilities that involve changes in the confining pressure both contain minus signs, to ensure that the numerical values of all four compressibilities will be positive. Many other compressibilities have been defined for porous rocks in the earth science and petroleum engineering literatures. Some alternative definitions and notations are reviewed by Chilingarian and Wolf [1975], Raghavan and Miller [1975], and Greenwald [1980].

In contrast to common engineering materials such as steels or ceramics, the various compressibilities of a porous rock vary with the state of stress, as well as with other parameters such as temperature. For sandstones, all four of these compressibilities typically decrease with increasing stress, and level off to constant values at confining pressures of about 40-80 MPa (6000-12,000 psi). The extent of nonlinearity is usually not negligible; differences of as much as a factor of ten between the high pressure and low pressure values of C_{bc} are not uncommon. Typical behavior of a porous rock compressibility as a function of stress is shown in Fig. 1.3, for a Frio sandstone from East Texas [Carpenter and Spencer, 1940]. This core was taken from a depth of 5460 feet, and had an initial porosity of 30%. Fig. 1.3a shows the pore strain as a function of confining pressure, with the pore pressure held constant at 0 psi. At low pressures the curve is very nonlinear, but above about 6000 psi the curve resembles that of a conventional elastic solid. The derived compressibility C_{pc} is shown in Figure 1.3b. The behavior of the other porous rock compressibilities is qualitatively similar to that shown for C_{pc} in Fig. 1.3.

The nonlinearity of the stress-strain curves of crystalline rocks were attributed by Adams and Williamson [1923] to the closure of very thin crack-like voids. This explanation carries over to sandstones, with the exception that the closable voids are the regions of imperfect bonding between grains. The numerical values of the porous rock compressibilities, as well as the variation of the compressibilities with stress, are to a large degree determined by the geometry of the void spaces in the rock. The values are also, of course, influenced by the compressibilities of the various mineral components of the sandstone. Much research has been

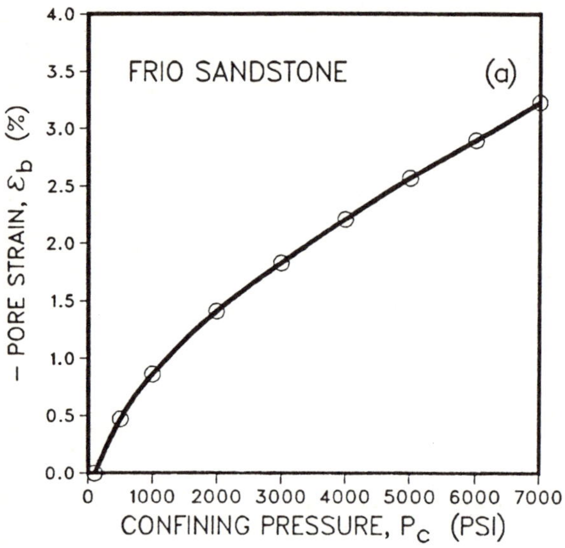

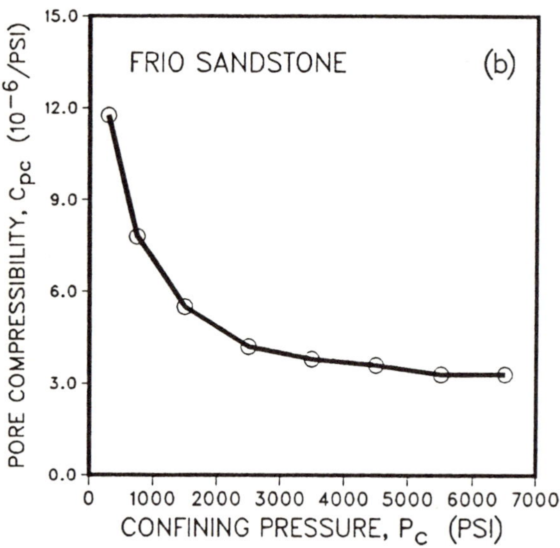

Fig. 1.3. Pore strain ε_p (a) and pore compressibility C_{pc} (b) in a Frio sandstone from East Texas, as a function of confining pressure, at zero pore pressure (after Carpenter and Spencer [1940]).

done to relate the numerical values of the compressibilities to mineralogy and pore structure. The state of knowledge in this area is discussed in detail in Part Two. Nevertheless, it is still generally true that compressibilities must be measured in the laboratory in order to arrive at values accurate enough for engineering calculations. As it is often not practical to measure all four compressibilities for each type of rock at all pressures of interest, it is desirable to have some method of correlating the different compressibility values to each other. This can be done using the theory of elasticity, along with a suitable idealized model for consolidated sandstones (see Chapter 2).

Although the derivatives appearing in eqns. (1.5-1.8) can be defined for any deformation process, the term "compressibility" is usually reserved for situations in which the deformation is elastic, which is to say *reversible*. In an elastic deformation, the strain at a given time is a single-valued function of the state of stress at that time, and does not depend on the past states of stress or strain. In particular, if stresses are removed from an elastic material, the strain will return to zero. Consolidated sandstones behave elastically for confining stresses that are less than some critical yield stress [see Zhang *et al.*, 1990], beyond which irreversible plastic deformation will occur. This plastic deformation is usually associated with crushing of the grains that comprise the sandstone, or with cleavage along the boundaries between grains. Another type of inelastic deformation is *fracture,* which can be caused by sufficiently high tensile stresses. A common situation in which this occurs is the process of *hydraulic fracturing,* in which fractures emanating from the borehole are induced by the injection of fluid into the borehole under high pressure [see Economides and Nolte, 1989]. Such modes of irreversible deformation lie outside of the scope of this work. Another type of deformation for which the compressibility coefficients do not determine the strains is the long-term compaction of sediments [Raghavan and Miller, 1975]. In these cases, deformation is is affected not only by the applied stresses, but also by non-mechanical effects such as cementation, pressure solution [Sprunt and Nur, 1977], and temperature [Schmoker and Gautier, 1989]. The above discussion should not be construed as implying that elastic deformation is an uncommon or specialized situation, however. Most of the deformation that occurs in petroleum or groundwater reservoirs as a result of fluid withdrawal is elastic, as is much of the deformation in crustal rocks due to tectonic forces.

Chapter 2. Relationships Between the Compressibilities

In order to develop a mathematical theory of the compressibility of porous rocks, it is first necessary to choose a conceptual model for rock-like materials. Once such a model is chosen, the laws of mechanics can be used to find relationships between the various compressibilities, and to gain some understanding of the manner in which these compressibilities vary with pressure. One of the earliest such models used for sedimentary rocks was that of a packing of spherical grains [Gassmann, 1951b; Brandt, 1955; Digby, 1981; Fig. 2.1a]. Models based on this idealization are particularly useful for unconsolidated sediments. The usefulness of these models does not entirely carry over to consolidated sediments, however. For example, while sphere-pack models account for the effect of confining pressure with some accuracy, it is difficult to treat pore pressure effects within this context. Another drawback to applying these models to consolidated rocks is that they typically predict compressibilities that decrease with pressure according to a power-law, while the compressibilities of consolidated sandstones actually level off to constant, non-zero values at high pressures.

A general theory of porous rock compressibility is more easily derived by starting with what is, in effect, the opposite model (see Fig. 2.1b). Instead of focusing on the grains, it is more convenient to consider a porous rock to be a solid containing discrete voids. More specifically, we will consider a sandstone to be composed of an isotropic, homogeneous elastic matrix, permeated with pores of various shapes and sizes. The matrix is assumed to form a completely connected network, while the void space may consist either of a connected network of pores, isolated pores, or some combination of both pore types. The connected portion of the pore space is often referred to as as the effective porosity, while the isolated voids are said to form the isolated or "noneffective" porosity [Clark, 1969, p. 20]. Although this distinction is crucial when studying transport properties such as permeability or electrical conductivity, it is not a necessary distinction to make when developing a general theory of the mechanical behavior of porous rocks. In the development that will be presented here, the voids need not be homogeneously distributed throughout the material, nor be random in their spatial orientation. Grain shape or size need not enter into consideration, as no explicit distinction will be made between the grain-forming minerals and any intergranular clays or cementing material. In this regard, it should be noted that the word "matrix" is used here to refer to the totality of the mineral components, and should not be confused with the use of that word by petroleum

(a)

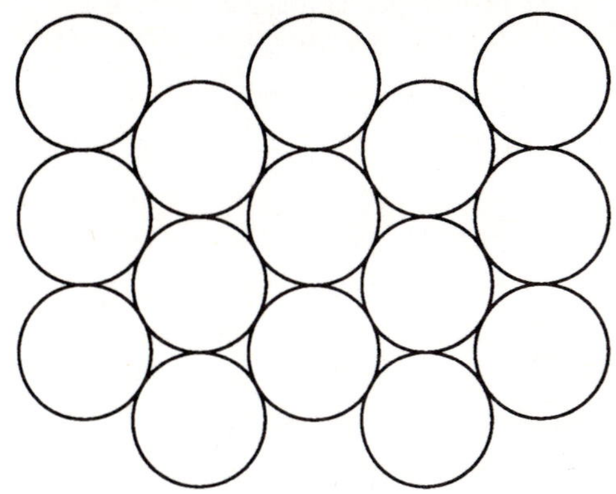

(b)

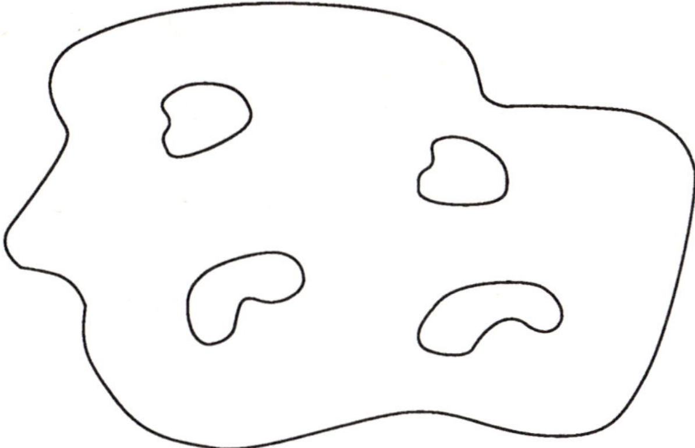

Fig. 2.1. (a) Sandstone modeled as an assemblage of discrete grains in contact; (b) modeled as a solid matrix containing voids.

geologists [Pettijohn, 1957, p. 284] to refer to certain types of intergranular material. Application of the laws of continuum mechanics to this model leads to a theory of porous rock compressibility that is quite general, in that it applies to both sedimentary and igneous rocks. In many respects it is useful for unconsolidated sediments, also. The extent to which any of the assumptions mentioned above are necessary for a particular theoretical result, and the effect that different assumptions might have, will be discussed as the results are derived.

For the idealized porous solid described above, the four porous rock compressibilities are not independent, and three relations can be found between them by using various concepts from elasticity theory. In the following derivation of these relations, applied pressures and the resulting strains will be *incremental* changes superimposed on a pre-existing state of stress and strain. While the stress-strain relations which result from this analysis will be nonlinear, representing the integration of incremental relations, the total strains will still be "infinitesimal" in the sense of classical linear elasticity [Sokolnikoff, 1956]. Higher order (*i.e.,* nonlinear) terms in the relationships between the strains and the deformation gradient [Murnaghan, 1951; Knopoff, 1963] will thus be neglected. For example, recall that the strain in the x direction is typically defined as $\varepsilon_x = \Delta L_x / L_x^i = \partial u_x / \partial x$, where u_x is the x–component of the displacement vector. The full nonlinear theory reveals that a more useful definition of the normal strain is $\varepsilon_x = \partial u_x / \partial x + (1/2)(\partial u_x / \partial x)^2$. The classical theory of linear elasticity neglects nonlinear terms such as $(1/2)(\partial u_x / \partial x)^2$. The validity of this approximation for application to the compression of sandstones is ensured by the fact that bulk strains never exceed a few percent for a consolidated rock in the elastic range of its behavior, since failure will occur at strains on the order of about 1-2% [Jaeger and Cook, 1979, pp. 144-153]. Pore strains may well be much larger than a few percent, particularly for low-porosity sandstones, but this causes no serious difficulty in applying the resulting theory, since the definition of pore strain ($d\varepsilon_p = dV_p / V_p^i$) does not require any restrictions on its magnitude.

The only type of boundary conditions which need be considered in the derivation of the compressibility relationships are uniform hydrostatic pressure of magnitude dP_c over the entire outer surface of the porous body, and uniform hydrostatic pressure of magnitude dP_p over the entire interior (pore) surface. This system of applied stresses will be denoted by $\{dP_c, dP_p\}$, where it is understood that either or both of the stress increments may be equal to zero. A key fact needed in this analysis is that [Geertsma, 1957] if a stress increment $\{dP, dP\}$ is applied to the surface of the body (that is to say, $dP_c = dP_p = dP$), the resulting incremental stress state in the rock is that of uniform hydrostatic pressure of magnitude dP throughout the matrix. This is verified by first noting that this state of stress satisfies the stress equilibrium equations [Sokolnikoff, 1956], which require that the divergence of the stress tensor vanishes. Since the divergence of a constant tensor is identically zero, this state of stress automatically satisfies stress equilibrium. Also, this state obviously satisfies the boundary conditions on both the outer and inner surfaces of the matrix. This stress state leads to a uniform isotropic dilatation of magnitude $d\varepsilon_r = -dP / K_r = -C_r dP$, where C_r is the compressibility of the rock matrix material, and $K_r = 1/C_r$ is its bulk modulus. The crucial step in this derivation is the recognition [Geertsma, 1957] that this state of stress and strain within the matrix is exactly the same as that which would occur if the pores were hypothetically filled up with matrix material, and the boundary conditions on the outer surface were left unchanged. In this latter case, the total bulk strain is equal to $-C_r dP$, so that the bulk volume change is given by $dV_b = -C_r V_b^i dP$.

Now consider the stress increment $\{dP, 0\}$, which corresponds to a change only in the confining pressure. By definition, this will give rise to a change in the bulk volume given by $dV_b = -C_{bc} V_b^i dP$. Since C_{bc} depends on the elastic moduli of the matrix material and the geometry of the pore space, it must in general be considered to be an additional independent parameter. If the stress increment is $\{0, dP\}$, the resulting change in bulk volume will, again by definition, be equal to $C_{bp} V_b^i dP$. For the infinitesimal changes being considered here, the principle of superposition is valid, so the stress increment $\{0, dP\}$ can be separated into the difference of the two increments $\{dP, dP\}$ and $\{dP, 0\}$ (see Fig. 2.2). The strains resulting from the stress increment $\{0, dP\}$ will therefore be equal to the difference between the strains resulting from the stress increments $\{dP, dP\}$ and $\{dP, 0\}$. Using the notation $dV_b(dP_c, dP_p)$ to refer to the bulk volume change resulting from the stress increment $\{dP_c, dP_p\}$,

$$dV_b(0, dP) = dV_b(dP, dP) - dV_b(dP, 0),$$

so

$$C_{bp} V_b^i dP = -C_r V_b^i dP - (-C_{bc} V_b^i dP) = (C_{bc} - C_r) V_b^i dP,$$

hence

$$C_{bp} = C_{bc} - C_r. \tag{2.1}$$

The principle of superposition is applicable since the compressibilities are nearly constant over the range of pressures from P to $P + dP$. Accounting for the variation of C_{bc} and C_{bp} with pressure would introduce terms of order $(dP)^2$ in the expressions for dV_b, which would become negligible compared to terms of order dP in the limit of infinitesimally small pressure increments.

A relation between C_{pc} and C_{pp} can be derived in a similar manner. Consider again the states of stress and strain which result from the application of the stress increment $\{dP, dP\}$, and recall that, in the matrix, this is exactly the same elastostatic state that would occur if the pores were filled in with matrix material and the body was subjected to a confining pressure dP. For this resulting hypothetical solid body, the local dilatation is again everywhere equal to

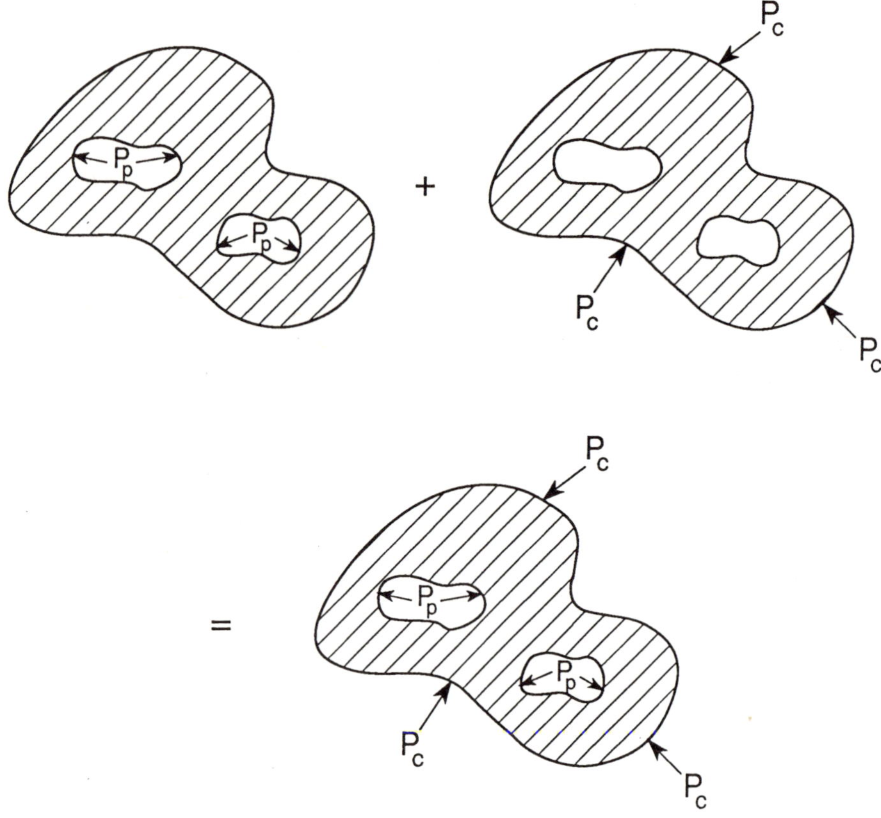

Fig. 2.2. Superposition argument used in the derivation of the relationships between the various compressibilities. If applied pressures are represented by $\{P_c, P_p\}$, then the state of strain due to $\{0, dP\}$ is equal to the difference between the states of strain due to $\{dP, dP\}$ and the state of strain due to $\{dP, 0\}$. Argument is valid only if the stress increment dP is small, so that the compressibilities can be considered constant.

$-C_r dP$. In particular, this is true for those regions of this solid body which had formerly been the pores of the porous body. Hence the dilatation of these regions of the porous body is also equal to $-C_r dP$, which implies that the total pore volume change is given by $dV_p = -C_r V_p^i dP$. Using the same stress state decomposition as before, in which $\{dP, dP\}$ = $\{dP, 0\}$ + $\{0, dP\}$, the changes in pore volume are

$$dV_p(0, dP) = dV_p(dP, dP) - dV_p(dP, 0),$$

so

$$C_{pp} V_p^i dP = -C_r V_p^i dP - (-C_{pc} V_p^i dP) = (C_{pc} - C_r) V_p^i dP,$$

hence

$$C_{pp} = C_{pc} - C_r. \tag{2.2}$$

The pore compressibilities and the bulk compressibilities can be related to each other through use of the Betti reciprocal theorem of elasticity [Sokolnikoff, 1956]. This theorem states that if an elastic body is acted upon by two sets of forces, say F_1 and F_2, the work done by F_1 acting through the displacements due to F_2 will exactly equal the work done by F_2 acting through the displacements due to F_1. Applying this theorem to the two sets of forces $\{dP, 0\}$ and $\{0, dP\}$, the work done by the first set acting through the displacements due to the second set is given by

$$W^{12} = -dP\,[dV_b(0, dP)] = -dP\,(C_{bp} V_b^i dP) = -C_{bp} V_b^i (dP)^2. \tag{2.3}$$

The minus sign is needed to account for the fact that a positive confining pressure will act in the direction opposite to the outward unit normal of the outer surface, while the bulk volume would increase if the displacement were in the direction of this normal. Similarly, W^{21} is given by

$$W^{21} = dP\,[dV_p(dP, 0)] = dP\,(-C_{pc} V_p^i dP) = -C_{pc} V_p^i (dP)^2. \tag{2.4}$$

No minus sign is needed in the defining equation for W^{21}, since the pore pressure acts in the same direction as that of increasing pore volume. The Betti reciprocal theorem implies that $W^{12} = W^{21}$, so that comparison of eqns. (2.3) and (2.4) reveals that $C_{bp} V_b^i = C_{pc} V_p^i$. Since the porosity ϕ is defined as V_p/V_b, this leads to the result that

$$C_{bp} = \phi^i C_{pc}. \tag{2.5}$$

It is worth noting that the use of the reciprocal theorem in this context does not depend on any assumptions concerning the geometry of the pore space. The reciprocal theorem is applicable to *any* two sets of forces applied to an elastic body, and in the present case one set of forces is considered to be the totality of the pore pressures applied to the entire pore surface of the rock. This use of the reciprocal theorem should not be confused with its use for the purposes of deriving explicit numerical values for the bulk volume change due to the pressurization of a single, isolated pore of a given shape, as is done, for example, by Walsh [1965a]. In this latter instance, explicit numerical results are obtained through the solution of the appropriate boundary value problem, which can only be accomplished if the pores are infinitely separated. Such attempts to relate the magnitude of the compressibilities to pore structure are the subjects of Part Two. In the present context, the displacements at the pore surface are related to those at the exterior surface, but explicit expressions for these displacements in terms of the shapes of the pores are not considered.

Eqns. (2.1), (2.2), and (2.5) provide three relations between the four compressibilities $\{C_{pc}, C_{pp}, C_{bc}, C_{bp}\}$, with ϕ^i and C_r as the only other parameters explicitly involved. Any three of the four porous rock compressibilities can therefore be expressed in terms of ϕ^i, C_r, and the other compressibility, although it is often convenient to consider C_{bc} to be the "fundamental" porous rock compressibility, since it is analogous to the compressibility of a non-porous material. In terms of C_{bc}, C_r and ϕ^i, the other three compressibilities are

$$C_{bp} = C_{bc} - C_r, \tag{2.6}$$

$$C_{pc} = (C_{bc} - C_r)/\phi^i, \tag{2.7}$$

$$C_{pp} = [C_{bc} - (1+\phi^i)C_r]/\phi^i. \tag{2.8}$$

It is worth noting that the equation $C_b = (1-\phi^i)C_r + \phi^i C_p$, which has often been presented without derivation [*i.e.,* Dobrynin, 1962], is not correct. Presumably, this expression is viewed as a "weighted average" of the compressibilities of the matrix and pore space. Aside from the fact that this equation does not distinguish between the effects of pore pressure and confining pressure, such a weighted average has no justification within the context of the present elastic analysis.

Since sandstones do not consist of an isotropic and homogeneous mineral phase, questions may be raised as to the appropriateness of the assumptions needed for the derivation of eqns. (2.6-2.8). On the length scale of distances between pores, the matrix material is usually a single grain of a relatively anisotropic crystal such as quartz, whose elastic moduli differ by as much as 40% along the different crystallographic axes [Clark, 1966]. Also, sandstones usually consist of more than one type of mineral. Common grain-forming minerals are [Pettijohn, 1957] quartz, feldspar, and calcite, while the intergranular spaces often contain clay minerals such as illite, kaolinite, or montmorillonite. Biot [1941] did not assume microscopic isotropy or homogeneity, and his analysis reproduces eqns. (2.1) and (2.5), but not eqn. (2.7). In Biot's theory, the difference between C_{pc} and C_{pp} is not necessarily equal to C_r, and the resulting equations contain an additional, independent parameter which can be identified [Brown and Korringa, 1975] with the change in pore volume caused by equal increments of both the pore and confining pressures.

Despite the failure of most rocks to physically conform to the idealized model discussed earlier, there are compelling reasons to use the results which follow from the assumption of matrix isotropy and homogeneity. For one, the calculations which are needed to relate compressibility to pore shape (see Part Two) become intractable without these two assumptions. Furthermore, the compressibilities of the major mineral components of most rocks do not differ by large amounts [see Clark, 1966; Simmons and Wang, 1971], so that the effective elastic moduli of the matrix are usually more well-behaved than the variegated mineralogical composition would at first imply. In fact, if the mineralogical composition of a given rock is known, it is a relatively simple matter (see below, this chapter) to compute a fairly accurate "effective" value for C_r, defined so that if the material were nonporous, a hydrostatic compression P would lead to an overall volumetric strain of $-C_r P$. Finally, the equations derived by Brown and Korringa [1975] for the more general case show that the overall bulk behavior is fairly insensitive to the numerical value of the additional Biot parameter.

While the theory of Biot [1941] is more general than that developed above, there have been many other treatments of porous rock behavior that include additional restrictive assumptions or simplifications, as opposed to greater generality. Often the volume change of the matrix material is neglected [Domenico, 1977], leading to equations which can be derived from eqns. (2.6-2.8) by setting C_r equal to zero. The two pore compressibilities are then equal to each other, as are the two bulk compressibilities. This assumption is acceptable for unconsolidated sands, for which C_{bc}/C_r may be as high as 10^2 [Newman, 1973]. For such materials, volumetric reduction occurs mainly in the vicinity of the grain contacts, as a result of the high contact stresses, as opposed to the hydrostatic compression of entire grains that occurs in consolidated rocks. An extensive treatment of the compressibility of unconsolidated sands is contained in the treatise edited by Chilingarian and Wolf [1975]. Most consolidated sandstones, however, will have a ratio of C_{bc}/C_r on the order of three or four, at least at effective pressures above about 10 MPa, in which case neglecting the compressibility of the rock minerals would not be a very accurate assumption.

The assumption that the mineral phase effectively behaves like an isotropic elastic medium can easily be tested experimentally through so-called "unjacketed tests". In these tests, which are discussed in more detail in Chapter 13, the rock is pressurized by a fluid which is allowed to seep into its pores, so that the pore and confining pressures are equal. Since $dP_p = dP_c$, eqn. (2.1) can be used to relate C_{bp} and C_{bc}, and so

$$d\varepsilon_b = -C_{bc}\,dP_c + C_{bp}\,dP_p$$

$$= -(C_{bc} - C_{bp})dP_c$$

$$= -C_r\,dP_c \ . \qquad (2.9)$$

Hence in an unjacketed test, the nonlinearity implicit in C_{bc} disappears, and the rock deforms as an elastic medium with compressibility C_r. An undrained compressibility test should therefore not only reveal whether or not the mineral phase behaves as an effective elastic medium, but should also allow a simple way of determining the value of C_r.

If the mineralogical composition of the rock is known, it is fairly easy to estimate a value for C_r by using the Voigt-Reuss-Hill method. Consider the problem of determining an effective C_r for a heterogeneous but non-porous body consisting of n different minerals, each with compressibility C_i, and present in a volume fraction of χ_i, arranged in any geometry. If such a body is subjected to uniform hydrostatic stress dP over its outer boundary, the state of stress in the various mineral grains will not be homogeneous, due to the necessity of maintaining continuity of both tractions and displacements at the grain boundaries. Reuss's method [see Hill, 1952; Zimmerman *et al.*, 1986b] for predicting the effective compressibility of such a body rests on assuming that the *stresses* are uniform throughout the body, and then ignoring the resulting discontinuity of displacements. If the stresses were uniform, the strain in each component would be $d\varepsilon_i = -C_i\,dP$. Since the total strain is the weighted average of the strains in each component,

$$\varepsilon = \Sigma\chi_i\varepsilon_i \ ,$$

$$d\varepsilon = \Sigma\chi_i\,d\varepsilon_i = -\Sigma\chi_i C_i\,dP \ ,$$

so

$$\bar{C}\,[\text{Reuss}] = -\frac{d\varepsilon}{dP} = \Sigma\chi_i C_i \ . \qquad (2.10)$$

In other words, the Reuss method consists of taking a volumetrically-weighted average of the individual compressibilities.

Voigt made the opposite assumption [see Hill, 1952; Zimmerman *et al.*, 1986b], which is that the *strains* are uniform throughout the body, so that in each grain $d\varepsilon = -C_i dP_i$. Since the average stress in the body is the volumetrically-weighted average of the stresses in each component,

$$P = \Sigma \chi_i P_i \ ,$$

$$dP = \Sigma \chi_i dP_i = -\Sigma(\chi_i / C_i) d\varepsilon \ ,$$

so

$$\bar{C}[\text{Voigt}] = -\frac{d\varepsilon}{dP} = \left[\Sigma(\chi_i / C_i)\right]^{-1} . \qquad (2.11)$$

The Voigt method therefore consists of taking a volumetrically-weighted average of the bulk moduli. Hill [1952] proved that the Voigt and Reuss estimates are actually upper and lower bounds of the true effective moduli, and proposed averaging these two estimates in order to find a best estimate of $\bar{C}$. Experimental data that justifies use of the Voigt-Reuss-Hill average for non-porous crystalline rocks has been compiled by Brace [1965]. The experimentally measured elastic moduli always lie between the Voigt and Reuss bounds. While there are cases in which the measured values are much closer to one of the bounds than they are to the Voigt-Reuss-Hill average, for most applications this average yields acceptable accuracy.

As an example of the use of the Voigt-Reuss-Hill method to estimate the effective compressibility, consider the three sandstones discussed by Greenwald [1980]. Table 2.1 lists the volume fractions of the various minerals in the three samples, and Table 2.2 lists their respective elastic moduli. As an example of the type of computation needed to estimate the effective compressibility, consider the Reuss estimate of the bulk modulus of the Bandera. According to eqn. (2.10) and Tables 2.1 and 2.2,

$$\bar{C}[\text{Reuss}] = \chi_Q C_Q + \chi_C C_C + \chi_F C_F$$

$$= \frac{\chi_Q}{K_Q} + \frac{\chi_C}{K_C} + \frac{\chi_F}{K_F}$$

Table 2.1

Mineralogical composition of three sandstones; data from Greenwald [1980].

Sandstone Mineralogical Compositions			
	Quartz	Calcite	Feldspar
Berea	.88	.05	.07
Boise	.64	.00	.36
Bandera	.70	.21	.09

Table 2.2

Elastic moduli of constituent minerals; data from Greenwald [1980].

Elastic Moduli of Minerals (units of 10^6 psi)			
	K	G	ν
Quartz	5.376	4.520	.1716
Calcite	9.755	4.975	.2821
Feldspar	9.145	4.545	.2868

Table 2.3

Effective elastic moduli of three sandstones, computed from eqns. (2.10) and (2.11).

Rock Matrix Elastic Moduli (units of 10^6 psi)			
	Berea	Boise	Bandera
G_{Voigt}	4.545	4.529	4.618
G_{Reuss}	4.543	4.529	4.611
G_{Hill}	4.544	4.529	4.614
K_{Voigt}	5.859	6.733	6.635
K_{Reuss}	5.667	6.313	6.188
K_{Hill}	5.763	6.523	6.412

Table 2.4

Hill effective elastic moduli of three sandstones, computed from eqns. (2.13) and (2.14).

Rock Matrix Elastic Moduli		
	C_r (units of 10^{-6}/psi)	ν_r
Berea	0.1753	.188
Boise	0.1533	.218
Bandera	0.1560	.210

$$= \frac{0.70}{5.376\times10^6\,\text{psi}} + \frac{0.21}{9.755\times10^6\,\text{psi}} + \frac{0.09}{9.145\times10^6\,\text{psi}}$$

$$= 0.1615\times10^{-6}/\text{psi}\ . \tag{2.12}$$

The Voigt estimate of the compressibility of Bandera is 0.1507×10^{-6}/psi. The Voigt-Reuss-Hill estimate is therefore 0.1561×10^{-6}/psi. The full set of Voigt, Reuss, and Hill estimates for the bulk and shear moduli K and G are shown in Table 2.3. When applying these methods to the determination of the effective shear modulus, it must be remembered that the compressibility is the multiplicative inverse of the bulk modulus. Hence, the proper analogy is between the bulk modulus and the shear modulus, not between the compressibility and the shear modulus. Hence, the Voigt estimate of the shear modulus is a volumetric average of the shear moduli of the mineral constituents, while the Reuss estimate is a volumetric average of the shear compliances, $1/G$.

Since experimental data is available on the unjacketed compressibility of Berea, consider the bulk moduli estimates for that rock. The Voigt estimate of 5.859×10^6 psi and the Reuss estimate of 5.667×10^6 psi are fairly close together. Since these estimates are upper and lower bounds, the actual effective compressibility must lie between them, notwithstanding uncertainties in the volume fractions and intrinsic compressibilities of the minerals. The Voigt-Reuss-Hill moduli K_H and G_H have been converted into compressibilities and Poisson's ratios in Table 2.4, using the formulas

$$C_r = \frac{1}{K_H}\ , \tag{2.13}$$

$$\nu_r = \frac{3K_H - 2G_H}{6K_H + 2G_H}\ . \tag{2.14}$$

The unjacketed compressibility measured by Anderson and Jones [1985] was 0.178×10^{-6}/psi. This compares extremely well with the calculated Voigt-Reuss-Hill estimate of 0.175×10^{-6}/psi. Note that the estimates discussed above can be applied to the prediction of the effective elastic moduli G_r and K_r, from which the Poisson's ratio can be found through eqn. (2.14). The weighted-averages of eqns. (2.10) and (2.11) cannot be applied directly to the calculation of ν_r; a weighted average of G and K does not lead to an analogous weighted average for ν. The Voigt and Reuss methods are also often applied, in elementary strength-of-materials texts, to the Young's modulus E. Although this procedure lends itself to suggestive "series" and "parallel" models, the Voigt and Reuss models lose their theoretical significance as upper and lower bounds when applied to moduli other than K or G.

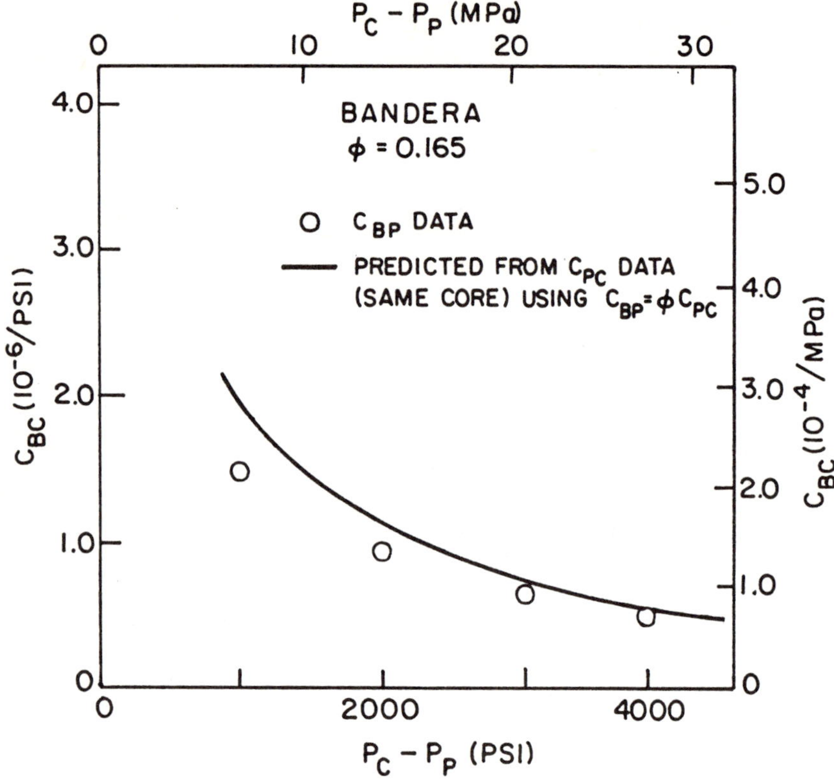

Fig. 2.3. Measured values [Zimmerman, 1984a] of C_{bp} for a Bandera sandstone, compared with the values predicted from C_{pc} measurements on the same core, using $C_{bp} = \phi^i C_{pc}$.

Verification of the compressibility relationships (2.6-2.8) is hampered by the fact that there are few published data sets in which more than one porous rock compressibility was measured on the same sandstone. Zimmerman [1984a] measured C_{bp} and C_{pc} for the same Bandera sandstone core. Bandera is a quartzose sandstone consisting of 70% quartzite, 21% calcite, and 9% feldspar [Greenwald, 1980]. The average grain diameter is about 80 μm, and the initial porosity of the sample was 16.5%. According to eqn. (2.5), C_{bp} should equal $\phi^i C_{pc}$. This relation follows from the reciprocal theorem of elasticity, and does not depend on the assumption that the mineral phase is isotropic or homogeneous. Fig. 2.3 shows the measured C_{bp} and C_{pc} values over the range of pressures from 0 to 30 MPa (0–4500 psi). Note that the compressibilities are plotted against the differential pressure $P_c - P_p$, which is shown in Chapter 5 to be the "effective stress" that governs the variation of the porous rock compressibilities. The agreement between C_{bp} and $\phi^i C_{pc}$ is fairly close, particularly in light of the relative inaccuracy of the "extruded volume" method that was used to measure C_{bp} (see Chapter 13).

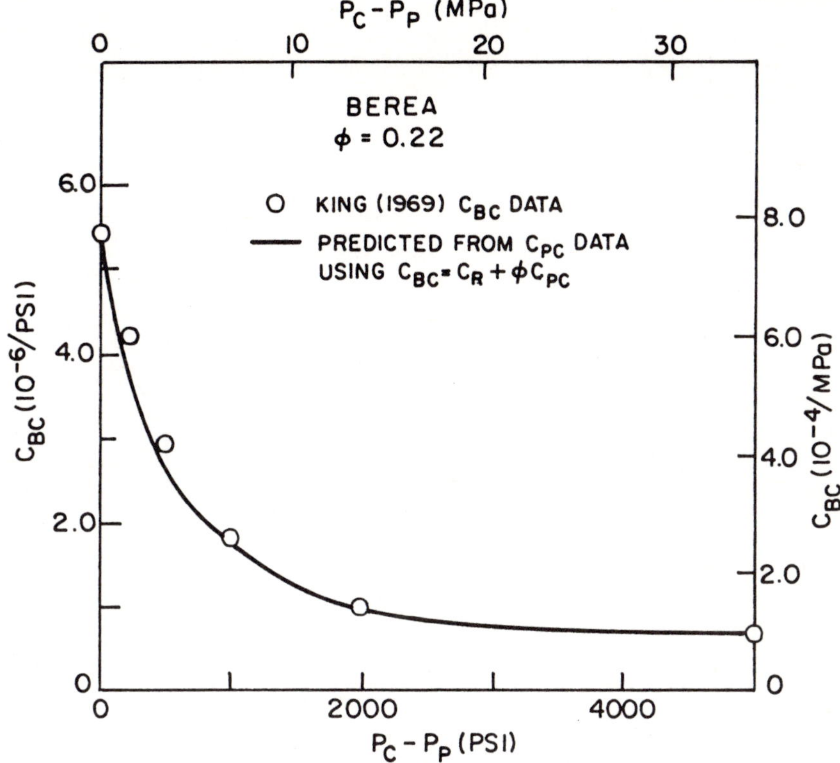

Fig. 2.4. Dynamically-measured values [King, 1969] of C_{bc} for a Berea sandstone, compared with the predictions based on Zimmerman's [1984a] C_{pc} measurements, using $C_{bc} = C_r + \phi^i C_{pc}$.

Another test of the compressibility relations (2.6-2.8) was described by Zimmerman [1984a]. C_{pc} pore compressibility measurements on a Berea sandstone of 22% porosity were converted into C_{bc} values using eqn. (2.7). The resulting $C_{bc}(P)$ curve was then compared to the values determined dynamically by King [1969] on a *dry* sample. The dynamic compressibility values were found by measuring the wavespeed of the longitudinal "P-wave", V_L, and the velocity of the transverse "S-wave", V_T. The relation $K = \rho(V_L^2 - 4V_T^2/3)$, where ρ is the bulk density of the rock, was then used to find the bulk modulus K, which equals $1/C_{bc}$. Although the tests were not conducted on the same specimens, each of the two Berea cores had an initial porosity of 22%. The excellent agreement (Fig. 2.4) not only verifies eqn. (2.7), but also illustrates that the dynamic compressibility of a dry sandstone is equivalent to the static value. This issue has often led to confusion, mostly caused by the fact that when acoustic waves travel through a *fluid-saturated* rock, the proper dynamic compressibility is *not* equal simply to C_{bc} (see Chapter 6).

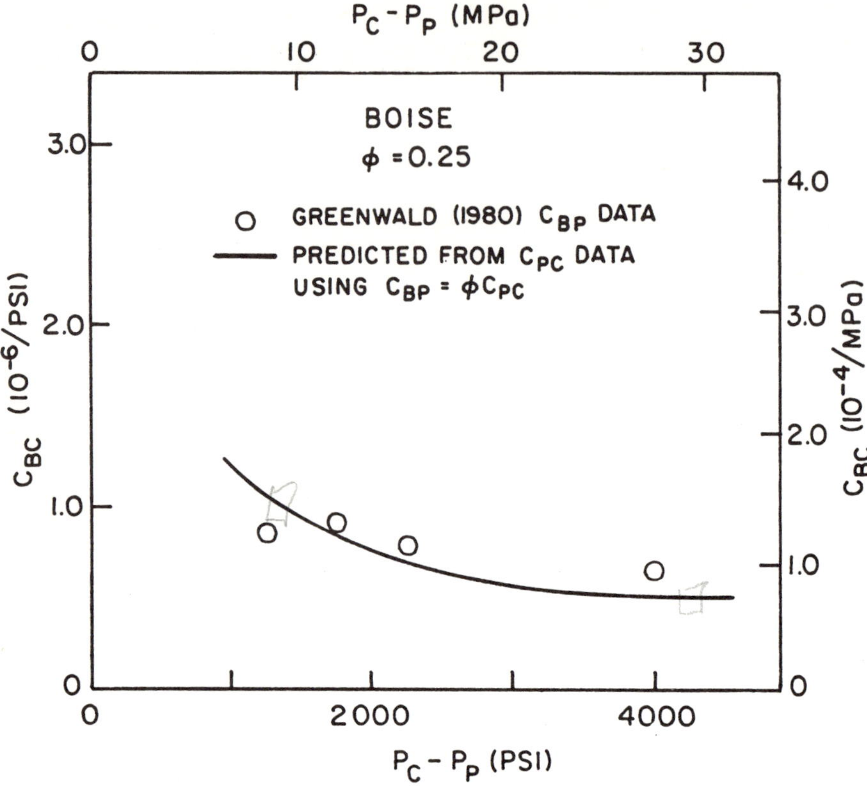

Fig. 2.5. Measured values [Greenwald, 1980] of C_{bp} for a Boise sandstone, compared with the predicted values based on Zimmerman's [1984a] C_{pc} measurements, using $C_{bp} = \phi^i C_{pc}$.

A similar verification of the compressibility relation between C_{bp} and C_{pc} is shown in Fig. 2.5 for a Boise sandstone of 25% initial porosity. Greenwald [1980] measured C_{bp} by varying the pore pressure and monitoring the resulting bulk volume change with strain gauges mounted on the outside of the rock. Zimmerman [1984a] determined C_{pc} by increasing the confining pressure on the rock, and measuring the volume of the expelled pore fluid. (These measurement techniques are discussed in more detail in Chapter 13.) The measurements were made on different cores that were cut from the same slab. According to eqn. (2.5), these two compressibilities should be related through $C_{bp} = \phi^i C_{pc}$. Fig. 2.5 shows the four discrete C_{bp} values measured by Greenwald, along with the C_{pc} values measured by Zimmerman, represented by a fitted curve. The agreement is fairly good, despite the inaccuracy inherent in compressibility data, which are found from ''numerically differentiating'' the stress-strain data. In particular, note that the average values of C_{bp} and $\phi^i C_{pc}$, over the range of 1000 – 4000 psi, are extremely close.

Chapter 3. Bounds on the Compressibilities

As mentioned in Chapter 2, the bulk compressibility C_{bc} of a porous rock plays a role analogous to the traditional compressibility C of a non-porous material. If sandstone were non-porous, C_{bc} would obviously be equal to the effective compressibility of the mineral phase, C_r. The pores and cracks in the rock have the effect of increasing the compressibility C_{bc} to a value greater than C_r. The precise amount of this increase depends on the porosity, as well as on the structure of the pores. However, within the context of the conceptual model of sandstone described in Chapter 2, lower bounds can be found on all four porous rock compressibilities, without any reference to the pore structure. These bounds can, in many cases, provide estimates of the compressibilities which are not too far from the measured values, and so they are of more than academic interest.

Consider a porous rock consisting of an homogeneous mineral phase permeated with pores that may be isolated or connected. The Voigt and Reuss methods (see Chapter 2) can be used to provide upper and lower bounds on the effective overall compressibility C_{bc} of such a rock. The mineral phase has compressibility C_r, and volume fraction $1-\phi$, while the "pore" phase has a volume fraction of ϕ and an effectively infinite compressibility. This analogy again corresponds to drained compression, in which the pore fluid is allowed to escape from the pore space as it contracts. Application of the Voigt and Reuss bounds (2.10) and (2.11) then immediately leads to

$$\frac{C_r}{1-\phi} \le C_{bc} < \infty \, . \tag{3.1}$$

Note that in this chapter, and much of the sequel, the porosity will be written without the subscript i.

According to these bounds, for example, the effective bulk compressibility of a sandstone with 20% porosity must be at least 25% greater than the compressibility of the mineral phase, regardless of the shapes and locations of the pores. Without any restrictions on the pore geometry, however, there is no finite upper bound on C_{bc}, since at any porosity there can be a connected network of pores and cracks that completely severs the rock and renders it incapable of resisting a load.

A somewhat higher, and therefore more useful, lower bound can be found by utilizing the results of Hashin and Shtrikman [1961], who used variational principles of elasticity theory to derive upper and lower bounds for the macroscopic elastic moduli of microscopically heterogeneous two-component materials. Application of the Hashin-Shtrikman bounds requires the additional restriction that the material is macroscopically isotropic, which implies that neither the pores nor the mineral grains have any preferential orientations. Since most sandstones are only mildly anisotropic, this is not too restrictive an assumption. (Measurements of sandstone anisotropy have been carried out by, among others, Wilhelmi and Somerton [1967], Mann and Fatt [1960], and Lo *et al.* [1986].) The Hashin-Shtrikman bounds can be applied to porous rocks if the matrix phase is taken as the first component, the pore space is taken as the second component, and C_{bc} is identified with the standard compressibility $C = 1/K$, leading to

$$0 \le \frac{K}{K_r} = \frac{C_r}{C_{bc}} \le \frac{2(1-2\nu_r)(1-\phi)}{2(1-2\nu_r) + (1+\nu_r)\phi} , \tag{3.2}$$

where ν_r is Poisson's ratio for the rock matrix material. As mentioned above, it is physically obvious that there can be no finite upper bound on the compressibility of a porous material, and indeed the Hashin-Shtrikman upper bound on C_{bc} is infinite. Since Hashin and Shtrikman [1961] showed that a space-filling packing of hollow spheres of various diameters, each with the same porosity ϕ, would have an effective bulk modulus equal to the upper bound in eqn. (3.2), this bound is obviously the sharpest one possible that does not depend on pore structure. Note that since $C = 1/K$, an upper bound on K/K_r is at the same time a lower bound for C_{bc}/C_r.

The Hashin-Shtrikman bounds can be combined with eqns. (2.6-2.8) to yield lower bounds on the other three compressibilities. This follows upon noticing that C_{bp}, C_{pc}, and C_{pp} are each monotonically increasing functions of C_{bc}, for fixed C_r and fixed ϕ. The absence of a finite upper bound on C_{bc} therefore implies the same lack of an upper bound for the other three porous rock compressibilities. In terms of the compressibility and Poisson's ratio of the rock matrix material, the four bounds on the porous rock compressibilities are

$$\frac{C_{bc}}{C_r} \ge 1 + \frac{3(1-\nu_r)\phi}{2(1-2\nu_r)(1-\phi)} , \tag{3.3}$$

$$\frac{C_{bp}}{C_r} \ge \frac{3(1-\nu_r)\phi}{2(1-2\nu_r)(1-\phi)} , \tag{3.4}$$

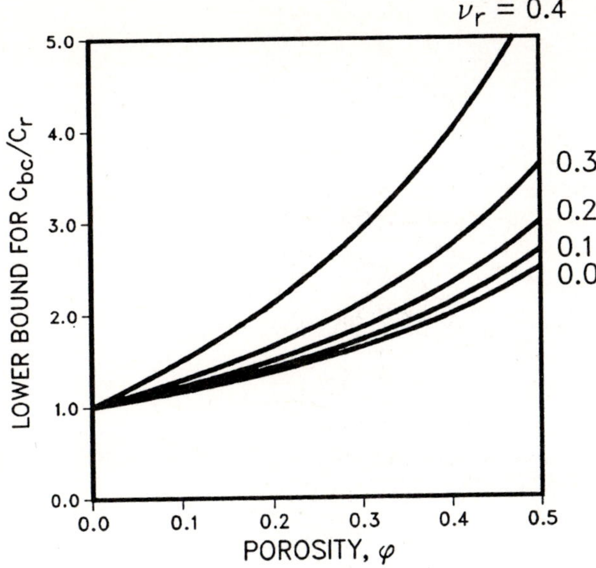

Fig. 3.1. Lower bound on C_{bc}, from eqn. (3.3).

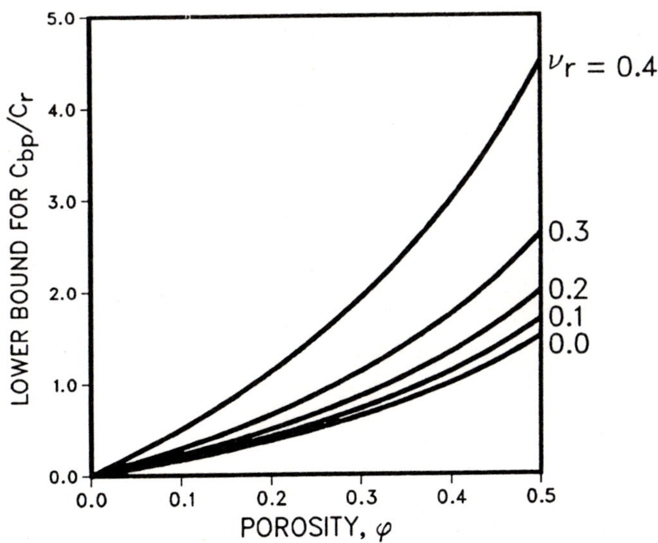

Fig. 3.2. Lower bound on C_{bp}, from eqn. (3.4).

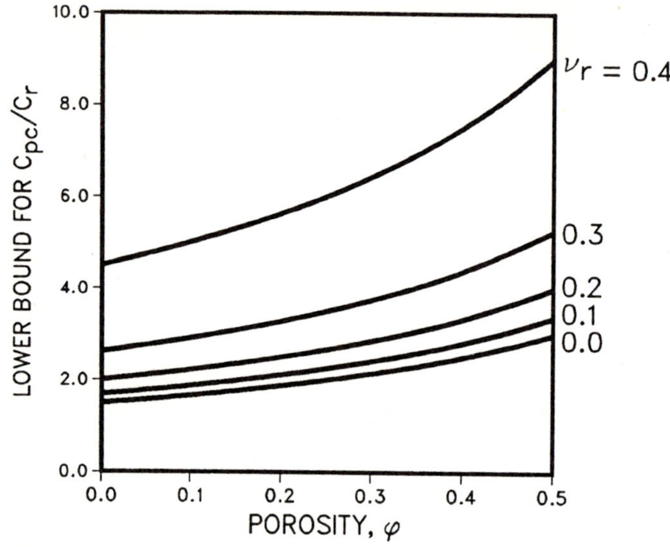

Fig. 3.3. Lower bound on C_{pc}, from eqn. (3.5).

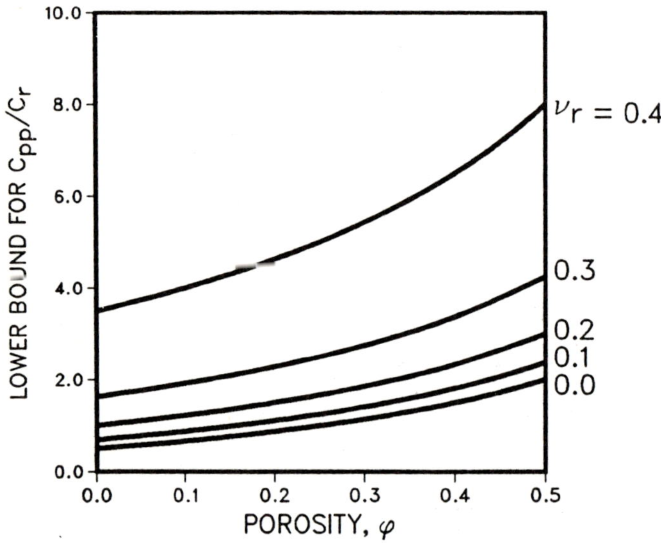

Fig. 3.4. Lower bound on C_{pp}, from eqn. (3.6).

$$\frac{C_{pc}}{C_r} \geq \frac{3(1-\nu_r)}{2(1-2\nu_r)(1-\phi)} \, , \tag{3.5}$$

$$\frac{C_{pp}}{C_r} \geq \frac{(1+\nu_r)+2(1-2\nu_r)\phi}{2(1-2\nu_r)(1-\phi)} \, . \tag{3.6}$$

These lower bounds are plotted in Figs. 3.1-3.4, for various values of ν_r, and for porosities up to 50%. This covers the entire range of porosities that would be of practical interest for application to sandstones, although the bounds are actually valid for all $0 \leq \phi < 1$. Each of the bounds is an increasing function of ϕ, and, when normalized against C_r, are also increasing functions of ν_r. This dependence of the lower bounds on ν_r is understandable in light of the fact that, for fixed C_r, the shear *stiffness* G_r is a *decreasing* function of ν_r. This can be seen by recalling that $G_r = 3(1-2\nu_r)/2C_r(1+\nu_r)$, from which it follows that $\partial G_r/\partial \nu_r = -9/2(1+\nu_r)^2 C_r < 0$. Since application of a hydrostatic confining pressure to the outer surface of a porous body leads to both a dilatation and a shear in the solid material, it is understandable that pore compressibilities will be affected by both the bulk and the shear modulus of the matrix material. Since the lower bounds are increasing functions of ν_r, bounds which do not depend on ν_r can be found by setting $\nu_r = 0$ in eqns. (3.3-3.6) (see Fig. 3.5):

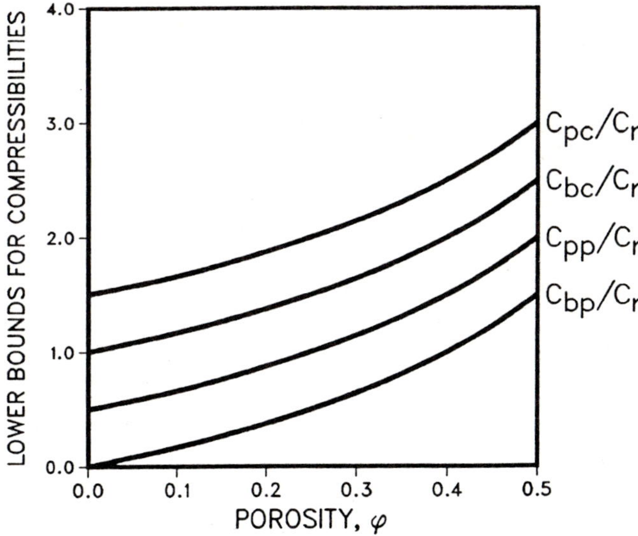

Fig. 3.5. Lower bounds on the porous rock compressibilities, from eqns. (3.7-3.10). These bounds are independent of the Poisson ratio of the matrix material.

$$\frac{C_{bc}}{C_r} \geq 1 + \frac{3\phi}{2(1-\phi)} , \tag{3.7}$$

$$\frac{C_{bp}}{C_r} \geq \frac{3\phi}{2(1-\phi)} , \tag{3.8}$$

$$\frac{C_{pc}}{C_r} \geq \frac{3}{2(1-\phi)} , \tag{3.9}$$

$$\frac{C_{pp}}{C_r} \geq \frac{1+2\phi}{2(1-\phi)} . \tag{3.10}$$

As mentioned above, the numerical values of the porous rock compressibilities will be strongly dependent upon the geometry of the pore space. It can be shown [Dewey, 1947; Mackenzie, 1950; Chapter 11] that for an elastic body containing a dilute dispersion of spherical pores,

$$\frac{C_{bc}}{C_r} = 1 + \frac{3(1-\nu_r)}{2(1-2\nu_r)}\phi . \tag{3.11}$$

This expression for C_{bc} corresponds exactly to the lower bound given by eqn. (3.3), in the limit of small ϕ. Hence, at least in the range of small porosities, it is known that spherical pores will cause the minimum possible increase in C_{bc}, as compared with pores of any other shape. For finite values of ϕ, computation of C_{bc} is a difficult problem which is as yet not completely solved, but it can reasonably be said that as pores deviate from sphericity, their effect on compressibility increases. The effect of pore shape on porous rock compressibilities is the subject of Part Two.

The stress-dependence of compressibility arises from the fact that as confining pressure is increased, the more compliant "crack-like" voids close up. Once closed, these voids no longer contribute to the compressibility, and so the compressibility decreases as the confining pressure increases. At "high" pressures, greater than about 40 − 60 MPa (6000 − 9000 psi), only the non-closable voids are still open. Whereas the crack-like voids are many orders of magnitude more compressible than the theoretical minimum given by eqn. (3.9), the non-closable voids are relatively less compressible. Hence the porous rock compressibilities typically start at very high values, and decrease towards their lower bounds at higher confining pressure. However, since the compressibilities of the non-closable voids are usually greater than the minimum values, the asymptotic high-pressure compressibilities lie above the lower bounds by a finite amount.

Table 3.1

Bulk compressibilities C_{bc} of twenty-four reservoir sandstones, as a function of the confining pressure, in units of 10^{-6}/psi. Pore pressure was held at 0 psi. Data are converted from the C_{pc} pore compressibility measurements of Fatt [1958a], using the equation $C_{bc} = C_r + \phi^i C_{pc}$. Note that Fatt's sample #13, a limestone, has been omitted from the table.

#	Source	Type	ϕ^i	0 psi	7000	15000
1	Brea Canyon, Ca.	Feldspathic graywacke	0.187	5.68	1.62	0.81
2	Huntington Beach, Ca.	Feldspathic graywacke	0.240	5.22	1.43	0.78
3	West Montalvo, Ca.	Feldspathic graywacke	0.099	3.74	1.02	0.78
4	West Montalvo, Ca.	Arkose	0.152	3.98	1.11	
5	Oxnard, Ca.	Lithic graywacke	0.090	4.69	1.57	
6	Oxnard, Ca.	Feldspathic graywacke	0.112	2.68	1.20	
7	Oxnard, Ca.	Graywacke	0.102	3.68	1.29	
8	Ventura, Ca.	Arkose	0.107	3.68	1.08	
9	San Joaquin Valley, Ca.	Arkose	0.176	2.54	0.68	0.58
10	San Joaquin Valley, Ca.	Feldspathic graywacke	0.214	4.18	1.20	0.70
11	San Joaquin Valley, Ca.	Feldspathic graywacke	0.199	7.00	1.83	0.98
12	San Joaquin Valley, Ca.	Lithic graywacke	0.089	1.77	0.58	0.38
14	Bradford, Pa.	Subgraywacke	0.145	1.18	0.46	
15	Sherman, Texas	Feldspathic quartzite	0.126	2.58	0.63	0.36
16	Kelly-Snyder, Texas	Feldspathic graywacke	0.169	2.18	0.72	0.38
17	Kelly-Snyder, Texas	Subgraywacke	0.192	1.71	0.51	0.38
18	Colorado	Feldspathic quartzite	0.211	2.21	0.64	0.37
19	Colorado	Subarkose	0.113	2.06	0.53	0.40
20	Wyoming	Feldspathic quartzite	0.144	2.68	0.65	
21	Wyoming	Orthoquartzite	0.093	2.23	0.47	
22	Nevada	Orthoquartzite	0.090	1.25	0.45	0.34
23	Louisiana	Arkose	0.241	4.18	0.89	0.38
24	Louisiana	Feldspathic graywacke	0.166	2.61	0.64	
25	South Ward, Texas	Feldspathic graywacke	0.118	3.16	0.95	

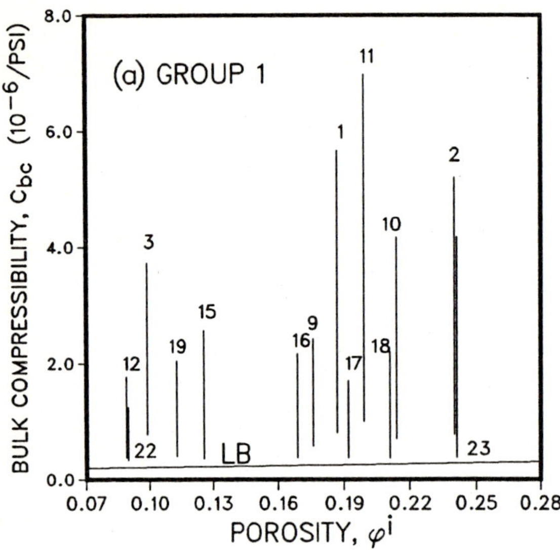

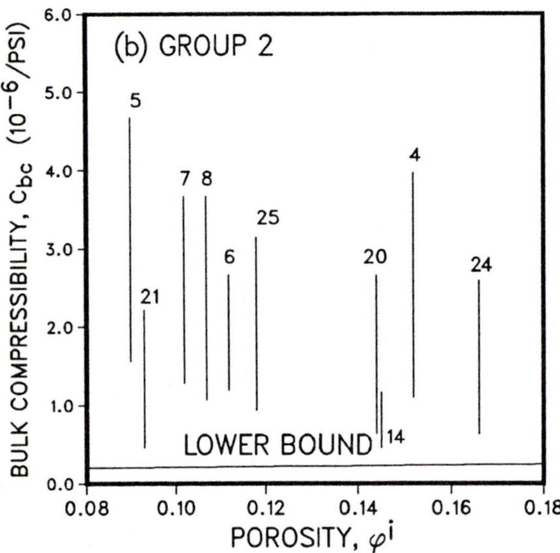

Fig. 3.6. Range of bulk compressibilities C_{bc} of twenty-four sandstones [after Fatt, 1958]. The upper end of each vertical line represents the value at $P_c = 0$, while the lower end represents the value at (a) 15,000 psi, or (b) 7000 psi. The numbers refer to the sample numbers, as listed in Table 3.1.

As an example of this behavior, consider the compressibilities measured by Fatt [1958a]. Fatt measured the pore compressibilities C_{pc} of twenty-four consolidated sandstones, and then converted these measurements to C_{bc} values through the (incorrect) equation $C_{bc} = \phi^i C_{pc}$. In Table 3.1 the corrected C_{bc} values are listed, by addition of the C_r term (see eqn. (2.7)), using Fatt's measured value of $C_r = 1.8 \times 10^{-7}$/psi. The sandstones were all petroleum-bearing rocks, and had initial porosities ranging from 8.9% to 24.1%. In fourteen cases ("group 1"), the compressibilities were measured over the range of confining pressures from 0–15,000 psi, while for the other ten cores ("group 2") the pressure range was 0–7000 psi. The pore pressures were held at 0 psi during these tests. The bulk compressibilities C_{bc} for the cores in group 1 are indicated in Fig. 3.6a, and those for group 2 are indicated in Fig. 3.6b, as functions of the initial porosity. The compressibilities for each core plot as vertical lines whose upper endpoints represent to compressibility at zero stress, and whose lower endpoints represent the high-stress asymptotes. The theoretical lower bound shown in the figures was calculated from eqn. (3.3) using a mineral phase compressibility of 1.8×10^{-7}/psi, and a Poisson ratio of 0.20. In each case, the compressibility starts at a high value at zero pressure, and decreases down towards the lower bound as the confining pressure increases. An interesting fact that also emerges from Fig. 3.6 is that while compressibility in general increases with porosity, there is certainly no simple relation between these two parameters. Correlations between compressibility and porosity that have been observed [Hall, 1953; Marek, 1971; Sampath, 1982] can be attributed to the small number of samples considered in each case.

Chapter 4. Effective Stress Coefficients

The concept of "effective stress" has long been used in rock mechanics. The motivation for this concept is that since the pore pressure and confining pressure tend to have opposite effects on the volumes (as well as on most petrophysical properties, such as permeability, electrical resistivity, etc.), it would be convenient to subtract off some fraction of the pore pressure from the confining pressure, and then treat the pressure as a single variable. In other words, instead of considering a property such as the pore volume to depend on the two variables $\{P_c, P_p\}$, it would be treated as a function of the single variable $P_c - nP_p$. This idea dates back to at least as far as the work of Terzaghi [1936], who proposed, based on a simple force-balance argument, that $P_e = P_c - \phi P_p$ would be the variable that governed the mechanical behavior of porous rocks. His experiments, however, revealed that the failure of geological materials was governed by the "differential pressure" $P_d = P_c - P_p$.

Following the introduction of the concept of effective stress, this concept has often been used in an overly simplified and imprecise way, by assuming that all properties of a porous rock could be expressed as functions of $P_e = P_c - nP_p$, where n is the "effective stress coefficient". More careful consideration of this matter has revealed that, insofar as this concept is valid, different effective stress coefficients must be used for the various mechanical deformation processes, such as pore volume compression, bulk volume compression, acoustic wave propagation, or failure [see Carroll and Katsube, 1983]. Furthermore, the concept has sometimes been used with reference to the variations of compressibility with pressure, and at other times used with reference to the changes in strain as a function of pressure; these two definitions are not, however, equivalent.

The most general form for expressing the effective stress concept for elastic porous rock deformation are the following expressions for the bulk and pore strains:

$$\varepsilon_b(P_c, P_p) = -C_{bc}(P_c - m_b P_p)[dP_c - n_b dP_p], \tag{4.1}$$

$$\varepsilon_p(P_c, P_p) = -C_{pc}(P_c - m_p P_p)[dP_c - n_p dP_p]. \tag{4.2}$$

In eqns. (4.1) and (4.2), the compressibilities C_{bc} and C_{pc} depend on the variable $P_c - mP_p$, and the strains are computed by multiplying the compressibilities by the bracketed stress increments. The effective stress coefficients m govern the manner in which the compressibilities vary with pressure, while the n coefficients reflect the relative amounts of additional strain caused by increments in the pore and confining pressures. Since the bracketed terms can always be written in that form by merely factoring out the appropriate compressibilities, the n coefficients can always be defined. However, there is no *a priori* reason to expect that the compressibilities will vary with stress as they do in eqns. (4.1) and (4.2), since in general a function of the two variables $\{P_c, P_p\}$ cannot be written as a function of some linear combination of the those two variables.

Comparison of eqns. (4.1) and (4.2) with eqns. (1.9) and (1.10) shows that the effective stress coefficients n are simply equal to the ratios of the appropriate compressibilities. The relationships (2.6-2.8) show that the effective stress coefficients n can be expressed as [Robin, 1973]

$$n_b = \frac{C_{bp}}{C_{bc}} = \frac{C_{bc} - C_r}{C_{bc}} = 1 - \frac{C_r}{C_{bc}}, \tag{4.3}$$

$$n_p = \frac{C_{pp}}{C_{pc}} = \frac{C_{bc} - C_r(1+\phi)}{C_{bc} - C_r} = 1 - \frac{\phi C_r}{C_{bc} - C_r} . \tag{4.4}$$

The effective stress coefficients are expressed in terms of C_{bc} in eqns. (4.3) and (4.4), although this is not necessary. The form of these equations shows that n_b and n_p will in general not be equal to each other.

Since the n coefficients depend on the compressibility C_{bc}, they will therefore vary with stress, as well as be dependent on pore structure and mineral composition. Various bounds and relationships can be found for these effective stress coefficients, however, thus greatly constraining their possibly range of values. C_{bc} and C_r are both positive, as is $C_{bc} - C_r$, so eqns. (4.3) and (4.4) imply that neither effective stress coefficient will exceed 1.0. This indicates that the pore and bulk volumes are each more sensitive to changes in confining pressure than to changes in pore pressure. Since n_b is a decreasing function of C_{bc}, the Hashin-Shtrikman bounds on C_{bc}/C_r lead immediately to the following bounds for n_b (Fig. 4.1), which are again independent of the pore geometry:

$$\frac{3(1-\nu_r)\phi}{2(1-2\nu_r)+(1+\nu_r)\phi} \le n_b \le 1. \tag{4.5}$$

The most general bounds, which are independent of ν_r, are [Norris, 1989]

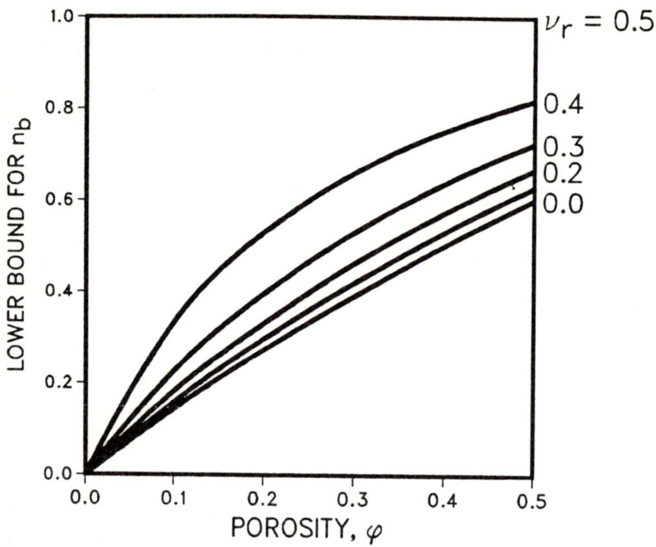

Fig. 4.1. Lower bounds on the effective stress coefficient n_b, from eqn. (4.5); the upper bound is equal to 1.0.

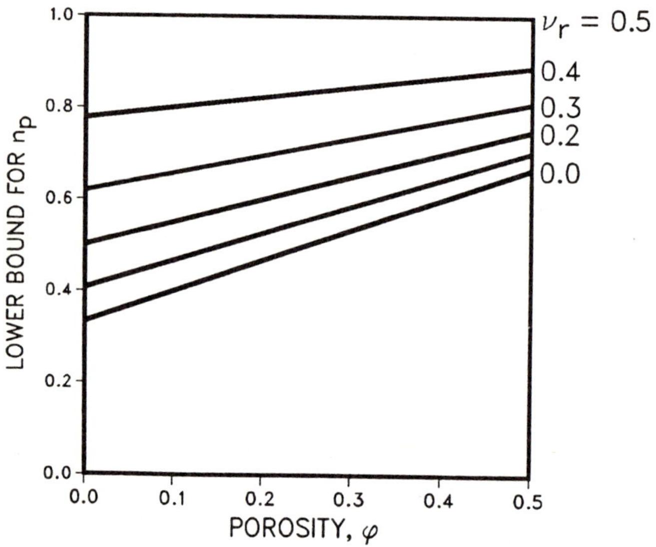

Fig. 4.2. Lower bounds on the effective stress coefficient n_p, from eqn. (4.8); the upper bound is equal to 1.0.

$$\frac{3\phi}{2+\phi} \leq n_b \leq 1 \,. \tag{4.6}$$

Since $3\phi/(2+\phi)$ always exceeds ϕ by a finite amount, n_b must be greater than ϕ [*cf.*, Knutson and Bohor, 1963]. As there is no upper bound on C_{bc}, either of the two n coefficients can approach arbitrarily close to unity.

Bounds on n_p can also be established, starting with the following expression that results from comparing eqns. (1.10) and (4.2):

$$n_p = \frac{C_{pp}}{C_{pc}} = \frac{C_{pc} - C_r}{C_{pc}} = 1 - \frac{C_r}{C_{pc}} \,. \tag{4.7}$$

In this form it is clear that the bounds for C_{pc} given in eqn. (3.5) lead to bounds for n_p (Fig. 4.2):

$$\frac{(1+\nu_r)+2(1-2\nu_r)\phi}{3(1-\nu_r)} \leq n_p \leq 1. \tag{4.8}$$

The most general lower bound, which again occurs for $\nu_r = 0$, is

$$\frac{(1+2\phi)}{3} \leq n_p \leq 1. \tag{4.9}$$

In the limit as ν_r approaches its maximum possible value of 0.50, which corresponds to an incompressible mineral phase, both n_b and n_p approach 1.0. Hence, if the matrix compressibility can truly be neglected, the pore and confining pressures will each have the same effect on the volumetric strains, except for sign.

The two coefficients n_p and n_b are not only in general different, but in fact n_p is always greater than n_b. This is proven by forming the difference $(n_p - n_b)$, and examining the sign of the resulting expression. In terms of C_{bc}, C_r, and ϕ,

$$n_p - n_b = \frac{C_r[C_{bc}(1-\phi) - C_r]}{C_{bc}(C_{bc} - C_r)}. \tag{4.10}$$

Since $(1-\phi)$ is the Voigt upper bound on C_r/C_{bc}, it follows that both the numerator and the denominator in eqn. (4.10) are non-negative. In fact, since the Voigt bound is never attained by a real porous material, because the Hashin-Shtrikman bound is more restrictive, the numerator must always be positive, proving that n_p will always be greater than n_b.

The most extensive investigations of the effective stress coefficients were those carried out by Fatt [1958a,1958b,1959]. Fatt measured both C_{bc} and C_{bp} for a Boise sandstone whose initial porosity was 26%. The Boise was a feldspathic graywacke consisting of 35% quartz, 29% feldspar, 6% rock fragments, 23% chert and clayey material, and 7% biotite [Fatt, 1958a]. The C_{bp} values were measured at a confining pressure of 12,000 psi, while the C_{bc} values were measured at zero pore pressure. Using a mineral phase compressibility of 2.0×10^{-7}/psi, which is roughly that of quartz or feldspar, Fatt compared the exact value of n_b, as given by C_{bp}/C_{bc}, with the theoretical value of $(C_{bc}-C_r)/C_{bc}$. C_{bp} was measured at a confining pressure of 12,000 psi, while C_{bc} was measured at a pore pressure of 0 psi. The results are shown in Fig. 4.3, as functions of the differential pressure $P_d=P_c-P_p$. The measured values of n_b were very close to unity for low values of the differential pressure, and decreased down to a high-stress asymptotic value of 0.77. The measured values agreed fairly well with the "theoretical" value, and each lied above the lower bound, which, for a Poisson ratio of 0.20, would be about 0.42.

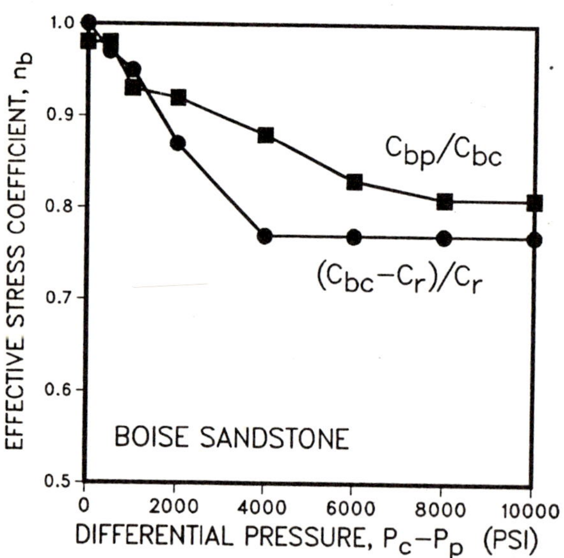

Fig. 4.3. Comparison of direct and indirect measurements [Fatt, 1958a] of n_b for a Boise sandstone.

The above discussion of effective stress coefficients utilized V_b and V_p as the basic kinematic variables. Another approach, which has been developed by Carroll and Katsube [1983], uses the "rock" volume V_r and the porosity ϕ as the basic variables. These variables are related to V_b and V_p by $V_r = V_b - V_p$, and $\phi = V_p/V_b$. This approach is somewhat less convenient for petroleum engineering purposes, since V_b and V_p are usually the more important parameters. However, the approach of Carroll and Katsube has the advantage of leading to extremely simple expressions for the effective stress coefficients. Furthermore, since it directly utilizes the solid rock volume V_r, this method is more readily generalized to inelastic behavior.

Starting with the definition $\phi = V_p/V_b$, and using the rules for differentiating quotients, an analysis similar to that carried out for V_p and V_b can be carried out for changes in the porosity:

$$\phi = V_p/V_b \, ,$$

so

$$\ln\phi = \ln V_p - \ln V_b \, .$$

Hence

$$\frac{d\phi}{\phi} = d\ln\phi = \frac{dV_p}{V_p} - \frac{dV_b}{V_b}$$

$$= C_{pp}\,dP_p - C_{pc}\,dP_c - C_{bp}\,dP_p + C_{bc}\,dP_c$$

$$= (C_{bc} - C_{pc})dP_c - (C_{bp} - C_{pp})dP_p$$

$$= \left[C_{bc} - \frac{[C_{bc} - C_r]}{\phi}\right]dP_c - \left[(C_{bc} - C_r) - \left\{\frac{[C_{bc} - C_r]}{\phi} - C_r\right\}\right]dP_p$$

$$= \left[C_{bc} - \frac{[C_{bc} - C_r]}{\phi}\right](dP_c - dP_p) \, .$$

Therefore,

$$d\phi = -[C_{bc}(1-\phi)-C_r]d(P_c - P_p) . \tag{4.11}$$

The effective stress coefficient n_ϕ is therefore always equal to 1.0, and the incremental porosity change depends only on the change in the difference between P_c and P_p. The term $[C_{bc}(1-\phi)-C_r]$ is necessarily positive, so the porosity is a *decreasing* function of the differential pressure $P_d = P_c - P_p$. This is a non-trivial conclusion since, for example, as an increase in confining pressure will decrease both the pore and bulk volumes, there is no *a priori* reason for the ratio V_p/V_b to also decrease in response to an increase in the confining pressure.

An equation for the strain of the rock matrix material can similarly be found by starting with the fact that $V_r = V_b - V_p$:

$$\frac{dV_r}{V_r} = \frac{dV_b - dV_p}{V_r} = \frac{1}{1-\phi} d\varepsilon_b - \frac{\phi}{1-\phi} d\varepsilon_p$$

$$= \frac{(C_{bp}\,dP_p - C_{bc}\,dP_c) - \phi\,(C_{pp}\,dP_p - C_{pc}\,dP_c)}{1-\phi}$$

$$= \frac{(C_{bp} - \phi C_{pp})dP_p - (C_{bc} - \phi C_{pc})dP_c}{1-\phi}$$

$$= \frac{(\phi C_{pc} - \phi C_{pp})dP_p \; - \; C_r dP_c}{1-\phi}$$

$$= \frac{\phi C_r dP_p \; - \; C_r dP_c}{1-\phi} ,$$

so

$$d\varepsilon_r = \frac{dV_r}{V_r} = \frac{-C_r d(P_c - \phi P_p)}{1-\phi} , \tag{4.12}$$

where the compressibility relations of Chapter 2 have been freely used. Hence the effective stress coefficient that governs changes in the volume of the mineral phase is simply equal to the porosity ϕ. Another interpretation of eqn. (4.12) is that the average stress $\bar{P}$ in the mineral phase of the rock is given by

$$\bar{P} = \frac{P_c - \phi P_p}{1-\phi} , \tag{4.13}$$

in the sense that $d\varepsilon_r = -C_r d\bar{P}$. This result can easily be motivated [Greenwald, 1980, p. 40] by the considering the two-dimensional force-balance argument shown in Fig. 4.4. If we assume that an "average" stress $\bar{P}$ acts over the interior rock material, equating the upward forces to the downward forces leads to the relationship $P_c = (1-\phi)\bar{P} + \phi P_p$, from which eqn.

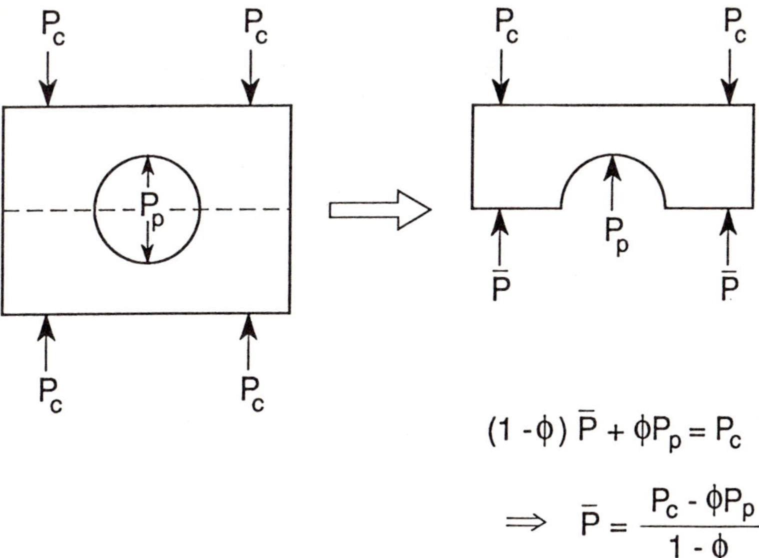

Fig. 4.4. Force-balance argument [after Greenwald, 1980] for the derivation of the mean-stress relationship. If the idealized rock on the left is cut by the dotted line (plane), the fractional amount of areal porosity that is exposed will be ϕ. The average stress acting over the exposed internal rock surfaces is denoted by $\bar{P}$. Equating the upward and downward forces then leads to eqn. (4.13).

(4.13) follows. This result can also be proven in a rigorous way, and furthermore can be shown to generalize to the regime of inelastic behavior [Carroll and Katsube, 1983]. Eqns. (4.11) and (4.12) show that by using V_r and ϕ as the kinematical variables, the effective stress coefficients n_r and n_ϕ are equal to 1.0 and ϕ, respectively. Hence, since ϕ usually will not vary by more than a few percent for an elastically deformed sandstone, these effective stress coefficients are essentially constant. Furthermore, since the rock volume compressibility C_r is also essentially constant, the nonlinearity of sandstone compression manifests itself only in the C_{bc} coefficient, in the Carroll-Katsube approach.

Chapter 5. Integrated Stress-Strain Relations

The stress-strain relations described by eqns. (1.9) and (1.10) refer to stress increments which are small enough so that the compressibilities can be considered constant throughout the process. These compressibilities are actually the local slopes of the stress-strain curves, and their multiplicative inverses are the so-called tangent moduli. They are useful for applications such as elastic wave propagation, in which the tangent moduli are the important parameters [Walsh, 1965b,1965c], since the stresses due to acoustic or seismic waves are small increments superimposed on an existing stress state. In other situations, such as subsidence calculations, the total strains are of more interest. Integration of eqns. (1.9) and (1.10) to yield the full stress-strain relationships requires knowledge of the precise variation of the compressibilities with both P_c and P_p, which leads to consideration of the m coefficients (see eqns. (1.1) and (1.2)).

Within the range of the elastic behavior of sandstones, it can be shown that the m coefficients must equal 1.0, which is to say that the compressibilities themselves are functions of the differential pressure $P_c - P_p$. Elastic behavior, strictly speaking, refers to situations where the state of strain at any point in time depends only on the state of stress at that point in time, and not on the past stress states, loading rates, etc. The term "elastic" in this context therefore does not necessarily imply "linear", although the term is often used with that implication in rock mechanics. If the rock behaves elastically, the two strain differentials (1.9) and (1.10) are mathematically "exact", which implies that the strains depend only on the values of P_c and P_p, and not on the specific details of the path taken in $\{P_c, P_p\}$ space. The path-independence of strain will not hold if there are frictional forces acting on the interfaces between grains, in which case there will be a hysteresis cycle. It has been argued by Walsh [1980], however, that these forces will not be present under hydrostatic loading that occurs in the absence of macroscopic shear.

Consider now a generalization of eqns. (4.1) and (4.2), in which C_{bc} and C_{pc} are considered to be arbitrary functions of P_c and P_p. Then, using eqns. (2.6) and (2.8) to eliminate C_{bp} and C_{pp},

$$d\varepsilon_b = -C_{bc}(P_c, P_p)dP_c + [C_{bc}(P_c, P_p) - C_r]dP_p, \tag{5.1}$$

$$d\varepsilon_p = -C_{pc}(P_c,P_p)dP_c + [C_{pc}(P_c,P_p)-C_r]dP_p. \tag{5.2}$$

The grain compressibility C_r can be considered to be a constant that is independent of pressure, since quantum mechanical arguments [Schreiber *et al.*, 1973, pp. 163-166], as well as extensive empirical data [Anderson *et al.*, 1968], show that the bulk compressibilities of rock-forming minerals do not change by more than a few percent over the range of confining pressures from zero up to about 100 MPa (about 15,000 psi), which would exceed the hydrostatic compressive strength of most sandstones. Furthermore, such stresses would exceed the maximum stresses that a sandstone would ever be subjected to *in situ*.

If a differential is exact, it must satisfy the Euler condition [Leighton, 1970, pp. 9-11], which in this context states that that if a single-valued function $\varepsilon(P_c,P_p)$ actually exists, its two mixed partial derivatives must be equal. Application of this test to the differentials $d\varepsilon_b$ and $d\varepsilon_p$ yields

$$\frac{\partial^2\varepsilon_b}{\partial P_p \partial P_c} = \frac{\partial}{\partial P_p}\left[\frac{\partial\varepsilon_b}{\partial P_c}\right] = \frac{-\partial C_{bc}}{\partial P_p},$$

$$\frac{\partial^2\varepsilon_b}{\partial P_c \partial P_p} = \frac{\partial}{\partial P_c}\left[\frac{\partial\varepsilon_b}{\partial P_p}\right] = \frac{\partial C_{bp}}{\partial P_c} = \frac{\partial(C_{bc}-C_r)}{\partial P_c} = \frac{\partial C_{bc}}{\partial P_c},$$

so

$$\frac{\partial C_{bc}}{\partial P_p} = -\frac{\partial C_{bc}}{\partial P_c}. \tag{5.3}$$

By applying the Euler criterion to ε_p, it can similarly be shown that

$$\frac{\partial C_{pc}}{\partial P_p} = -\frac{\partial C_{pc}}{\partial P_c}. \tag{5.4}$$

Both C_{bc} and C_{pc} must therefore satisfy the same simple partial differential equation, $\partial f(x,y)/\partial x = -\partial f(x,y)/\partial y$, whose general solution is $f(x,y) = f^*(x-y)$, where f^* is *any* function which depends on x and y only through the combination $(x-y)$. The exactness condition on the strain differentials therefore implies that the compressibilities depend on the pressures only through the differential pressure $P_d = P_c - P_p$. Thus

$$C_{bc} = C_{bc}(P_c - P_p), \tag{5.5}$$

$$C_{pc} = C_{pc}(P_c - P_p), \tag{5.6}$$

which is to say that $m_b = m_p = 1.0$. Relations (2.1) and (2.2) then show that C_{bp} and C_{pp} also depend only on the differential pressure.

These results can also be understood in terms of the following physical considerations. The porous rock compressibilities will change only if there is a qualitative alteration of the geometry of the pore space. Pores which are more or less ''equi-dimensional'' will not undergo drastic changes in shape, since their strains will never exceed a few percent in the elastic range of deformation. Void spaces of small aspect ratio, such as the regions of imperfect contact between grains, will, however, close up completely under sufficiently high confining stress, at which point they cease to contribute to the overall compressibility of the rock. (Closed cracks would still affect the behavior of the rock under *deviatoric* loading, *i.e.*, shear stresses.) Consider, for example, an isolated crack which is in the shape of an oblate spheroid whose minor and major axes are in the ratio of α:1, where α is the aspect ratio. It can be shown, using the results of Sneddon [1946] and superposition arguments similar to those used in Chapter 2, that the compression of such a crack is described by (see Chapter 10)

$$\varepsilon[\text{crack}] = \frac{\Delta V_p}{V_p^i} = \frac{-4(1-\nu_r^2)}{3\pi(1-2\nu_r)\alpha} C_r (P_c - P_p) - C_r P_p \ . \tag{5.7}$$

Since the elastic moduli of any rock-forming mineral will be on the order of 100 GPa (see Table 2.2), the aspect ratios of cracks which can be completely closed by pressures below about 100 MPa (15,000 psi) must be much less than 0.01; this is seen from eqn. (5.7) by letting $\varepsilon[\text{crack}] \rightarrow -1$. The first term on the right in eqn. (5.7) will therefore be much larger than the second term for closable cracks, since the second term does not include the factor $1/\alpha$. Hence, the closure of the crack will be governed, to a very good approximation, by the differential pressure $P_d = P_c - P_p$. Crack models other than the spheroid [Walsh, 1965a; Mavko and Nur, 1978; see also Chapter 9] merely lead to slightly different forms for the terms which depend on ν_r, but do not alter the conclusion that the differential pressure dominates crack closure, and thereby determines the variation of compressibility with stress.

A systematic and convincing demonstration that the m coefficients of consolidated sandstones are in fact equal to 1.0 has been given by Zimmerman *et al.* [1986a]. Pore strains were measured on three sandstones (Boise, Berea, and Bandera; see description in Chapter 2) as a function of the confining stress, with the pore pressure held constant. Such tests were carried out at various values of the pore pressure. At each pore pressure, the pore strain was fitted to the confining pressure with a curve of the form

$$\varepsilon_p = AP_c - Be^{-P_c/\hat{P}} + D \ , \tag{5.8}$$

where A, B, D and $\hat{P}$ are constants which are found by a least-squares regression. Since $\hat{P}$ enters eqn. (5.8) in a nonlinear fashion, this is not a simple linear regression problem. A, B and D were found for fixed values of $\hat{P}$, after which an optimum value of $\hat{P}$ was found by minimizing the variance between the measured strains and the fitted curve. In each case, the standard deviation between the optimum curve and the data was no more than about 1%, and the correlation coefficients were all greater than 0.9990. The pore compressibility C_{pc} is then found by differentiating eqn. (5.8) with respect to P_c:

$$C_{pc} = A + (B/\hat{P})e^{-P_c/\hat{P}} \ . \tag{5.9}$$

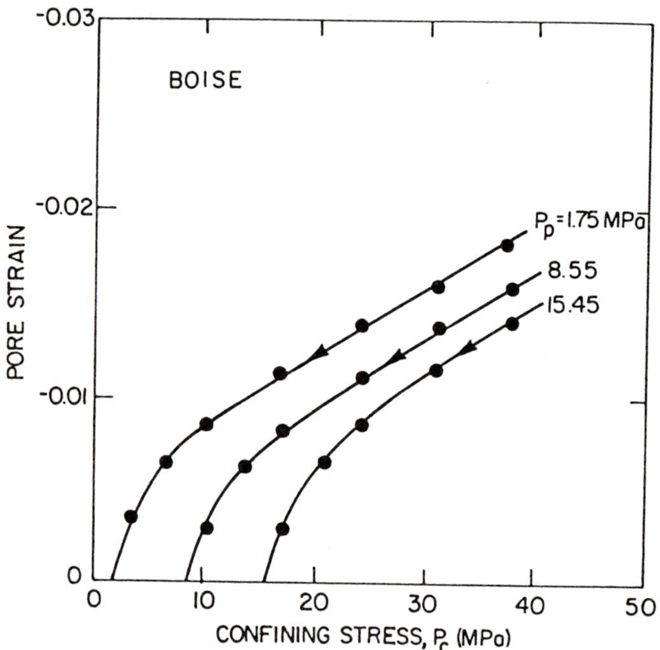

Fig. 5.1. Pore strain of a Boise sandstone as a function of the confining pressure, at three different pore pressures [Zimmerman *et al.*, 1986a]. Curves are found from least-squares fits to functions of the form given in eqn. (5.8). Since only *changes* in the pore strain were measured, the curves and the data points have been shifted so as to vanish when the differential pressure is zero. This was done for purposes of comparison, and has no effect on the computed compressibilities.

The pore strain curves for a Boise sandstone of 25.6% porosity are shown in Fig. 5.1 as functions of the confining pressure, for three different pore pressures. Since the strains were only measured to within an additive constant, the curves have been shifted vertically so that the pore strains are all zero when $P_c = P_p$. This has no effect on the computed compressibilities, which are shown in Fig. 5.2, since the constant D in eqn. (5.8) does not appear in the compressibility equation (5.9). The Boise sandstone was typical of the other two sandstones insofar as the degree to which the data were fit by the curves. If the effective stress concept is correct, the curves should all more or less coincide if they are shifted to the left by an amount $m_p P_p$. The best estimate of m_p was found by minimizing the standard deviation of the three compressibility curves, at ten equally spaced effective pressures from 3.45 to 34.5 MPa (500 – 5000 psi). It was found that the results were not affected by altering the number of pressures considered, or by ignoring the higher or lower extremes of the pressure range. The optimal value of m_p was found to be 1.02, for which the average standard deviation of the three curves at these ten pressures was 0.02650. Using the value m_p = 1.0 increases this average deviation by only 2%, to 0.02702, while a value such as m_p = 0.90 increases this measure of scatter by 80%, to 0.04832. It is therefore clear that the optimal value of m_p is is very close to unity, with a high probability. In Fig. 5.3, the compressibilities are plotted as functions of the differential pressure $P_c - P_p$, and it is seen that the curves show much less scatter than those in Fig. 5.2.

The results of a similar analysis of pore compressibility measurements on a Bandera (ϕ = 16.5%) and a Berea (ϕ = 22.2%) sandstone are shown in Figs. 5.4 – 5.7. The pore compressibilities are plotted as functions of P_c in Figs. 5.4 and 5.6, and as functions of the differential pressure in Figs. 5.5 and 5.7. The optimal value of m_p was found to be 1.06 for the Bandera, and 1.02 for the Berea. Note that the apparent breakdown of the effective stress law at low pressures is to some extent an artifact of fitting the compressibilities to an exponential equation such as eqn. (5.9). If the stress-strain curve fit is performed manually, and the curve "differentiated" with a protractor, the compressibility curves show much less scatter at low values of the differential pressure [Zimmerman *et al.*, 1985].

If the m coefficients are assumed to equal 1.0, as implied by the above arguments, the strain differentials given in eqns. (4.1) and (4.2) can be integrated along any path in $\{P_c, P_p\}$ space. A simple, and physically meaningful, path is that along which the confining pressure is first increased from zero to its final value, while the pore pressure remains at zero, after which P_c remains fixed while P_p increases to its final value (Fig. 5.8). (Note that almost all laboratory measurements of physical properties of reservoir rocks have been conducted under *positive* differential pressures, *i.e.*, $P_c > P_p$, which fall below the 45° line in Fig. 5.8. One exception is the work of Shankland and Halleck [1981].) For the first stage of this path, $P_p = 0$, so $(P_c - P_p) = P_c$, and $d\varepsilon_b = -C_{bc}(P_c)dP_c$. Using P without a subscript as a dummy variable of integration,

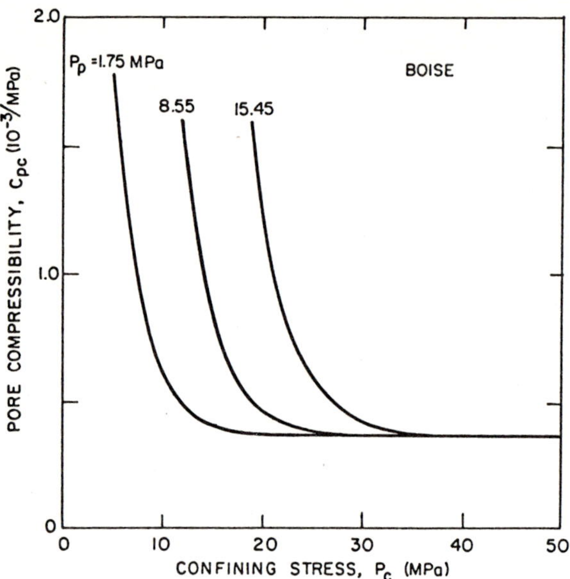

Fig. 5.2. Pore compressibility C_{pc} of a Boise sandstone as a function of confining pressure, at three different pore pressures [Zimmerman *et al.*, 1986a]. Compressibilities are found by differentiating the curves that were fit through the stress-strain data, such as those shown in Fig. 5.1.

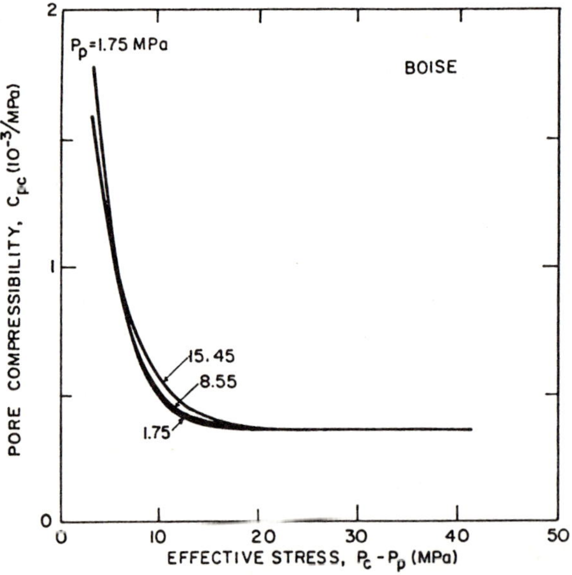

Fig. 5.3. Same data as in Fig. 5.2, plotted as a function of $(P_c - m_p P_p)$, with $m_p = 1$. The optimum value of m_p was actually 1.02.

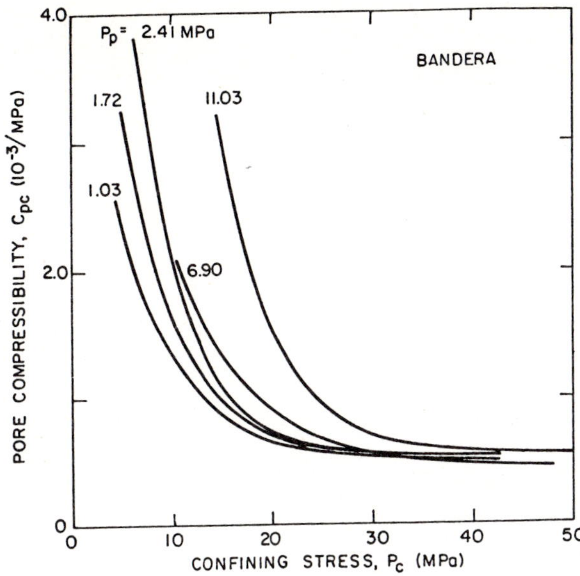

Fig. 5.4. Pore compressibility of a Bandera sandstone as a function of confining pressure [Zimmerman *et al.*, 1986a].

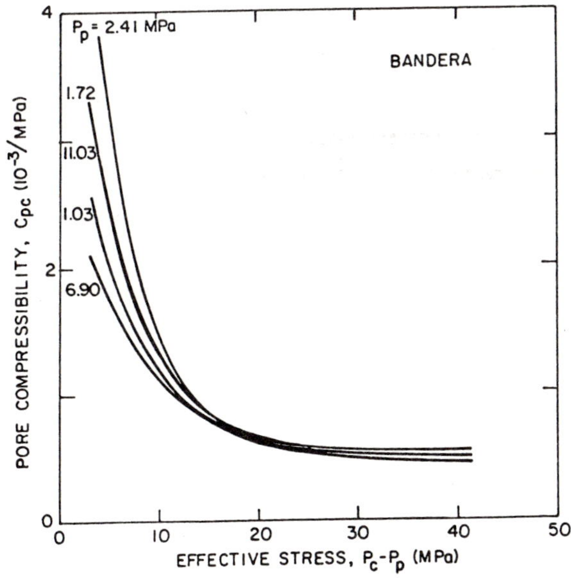

Fig. 5.5. Same data as in Fig. 5.4, plotted as a function of $(P_c - m_p P_p)$, with $m_p = 1$. The optimum value of m_p was actually 1.06.

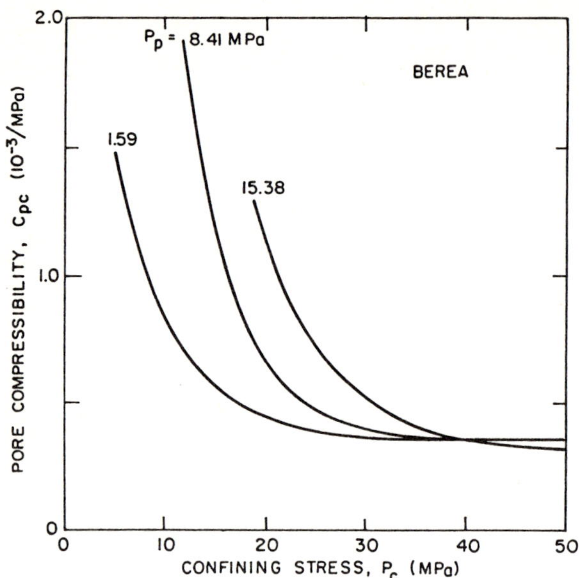

Fig. 5.6. Pore compressibility of a Berea sandstone as a function of confining pressure [Zimmerman *et al.*, 1986a].

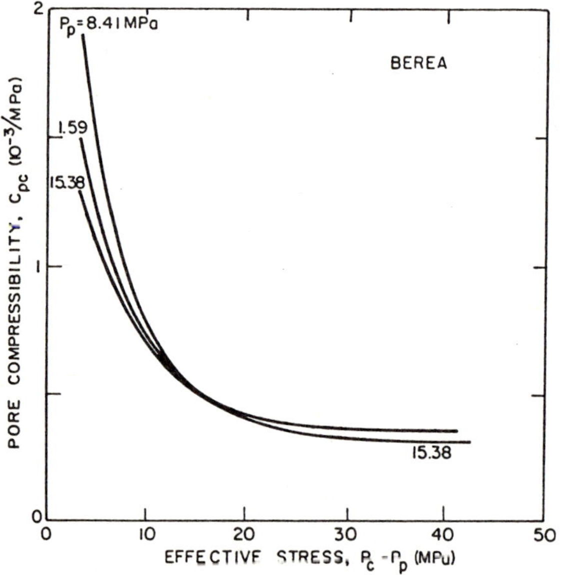

Fig. 5.7. Same data as in Fig. 5.6, plotted as a function of $(P_c - m_p P_p)$, with $m_p = 1$. The optimum value of m_p was actually 1.02.

$$\varepsilon_b^1 = \int d\varepsilon_b = -\int_0^{P_c} C_{bc}(P)dP \ . \tag{5.10}$$

For the second stage of this path, P_c is fixed, so $d\varepsilon_b = C_{bp}dP_p$, and, using eqn. (2.6),

$$\varepsilon_b^2 = \int d\varepsilon_b = \int_0^{P_p} [C_{bc}(P_c - P) - C_r]dP \ . \tag{5.11}$$

Putting $P' = P_c - P$, then $dP' = -dP$, and this integral can be taken from $P' = P_c$ to $P' = (P_c - P_p)$, as follows:

$$\varepsilon_b^2 = \int d\varepsilon_b = -\int_{P_c}^{(P_c - P_p)} [C_{bc}(P') - C_r]dP'$$

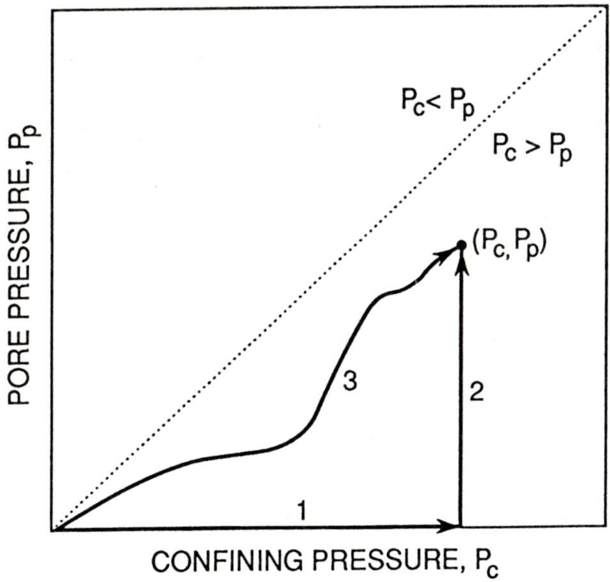

Fig. 5.8. Stress paths in $\{P_c, P_p\}$ space. Paths 1 and 2 are used in text in the derivation of the integrated stress-strain relations (5.13) and (5.14). Path 3 is a generic path from {0,0} to $\{P_c, P_p\}$.

$$= -\int_{P_c}^{(P_c-P_p)} C_{bc}(P')dP' - C_r P_p \,. \tag{5.12}$$

The total strain, in light of the fact that the strain is zero in the stress-free state, is given by the sum of the terms ε_b^1 and ε_b^2:

$$\varepsilon_b(P_c,P_p) = -\int_0^{P_c} C_{bc}(P)dP - \int_{P_c}^{(P_c-P_p)} C_{bc}(P)dP - C_r P_p$$

$$= -\int_0^{(P_c-P_p)} C_{bc}(P)dP - C_r P_p \,. \tag{5.13}$$

The equivalent expression for ε_p is [Chierici *et al.*, 1967]

$$\varepsilon_p(P_c,P_p) = -\int_0^{(P_c-P_p)} C_{pc}(P)dP - C_r P_p \,. \tag{5.14}$$

In these forms, each strain depends on one term (the integral) that involves the differential pressure P_d, and a second term which involves only the pore pressure P_p. At first sight it does not seem as if the effective stress coefficients n_b and n_p will be useful, since they do not explicitly appear in eqns. (5.13) or (5.14). Consider now, though, a process in which P_p remains fixed at zero, whence

$$\varepsilon_b = -\int_0^{P_c} C_{bc}(P)dP. \tag{5.15}$$

The "secant compressibility" $\overline{C}_{bc}$ can be defined (Fig. 5.9) to be the ratio of $-\varepsilon_b$ to P_c:

$$\overline{C}_{bc} = -\frac{\varepsilon_b}{P_c} = \frac{1}{P_c}\int_0^{P_c} C_{bc}(P)dP. \tag{5.16}$$

$\overline{C}_{bc}$ can be interpreted as the average value of C_{bc} over the range of pressures from 0 to P_c. In terms of the differential pressure $P_d = P_c - P_p$, the general expression given by eqn. (5.13) can be written as

$$\varepsilon_b(P_c,P_p) = -\overline{C}_{bc}(P_d)P_d - C_r P_p , \tag{5.17}$$

in which $\overline{C}_{bc}(P)$ is first evaluated at $P = P_d$, and is then multiplied by P_d. This can be written in an "effective stress" form as follows:

$$\varepsilon_b(P_c, P_p) = -\overline{C}_{bc}(P_d)\left[P_d + \frac{C_r}{\overline{C}_{bc}(P_d)}P_p\right]$$

$$= -\overline{C}_{bc}(P_d)\left[P_c - P_p + \frac{C_r}{\overline{C}_{bc}(P_d)}P_p\right]$$

$$= -\overline{C}_{bc}(P_d)\left[P_c - \left\{1 - \frac{C_r}{\overline{C}_{bc}(P_d)}\right\}P_p\right]$$

$$= -\overline{C}_{bc}(P_d)\left[P_c - \overline{n}_b P_p\right], \tag{5.18}$$

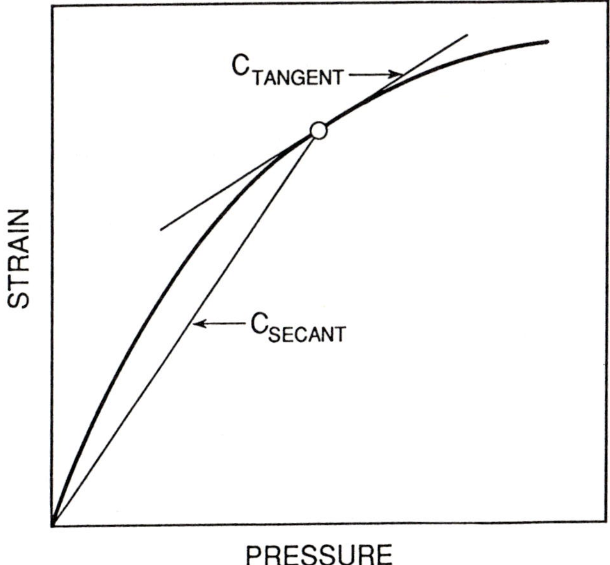

Fig. 5.9. Schematic depiction of the difference between the tangent and secant compressibilities. $C_{\tan}$ measures the local slope of the stress-strain curve at pressure P, while $C_{\sec} = \overline{C}$ measures the average slope of the curve from 0 to P.

where $\bar{n}_b$ is a sort of "secant value" of n_b. Since $\bar{C}_{bc}$ is an average of C_{bc}, the Hashin-Shtrikman bounds apply to it, and so the associated bounds on n_b are applicable to $\bar{n}_b$. The analogous expression for ε_p is [Zimmerman *et al.*, 1986a]

$$\varepsilon_p(P_c,P_p) = -\bar{C}_{pc}(P_d)\Big[P_c - \bar{n}_p P_p\Big], \tag{5.19}$$

where

$$\bar{n}_p = 1 - \frac{C_r}{\bar{C}_{pc}(P_d)}. \tag{5.20}$$

It should be noted that for low porosity rocks, eqn. (2.7) cannot be indiscriminately used to express the pore strain integrals in terms of C_{bc} instead of C_{pc}, without accounting for the fact that the porosity varies with stress during the integrations. This is not, however, of major concern for most sandstones, for which ϕ can be assumed "constant" as far as eqns. (2.6-2.8) are concerned.

It is therefore seen that for neither the pore strain nor bulk strain can the total strain be expressed purely in terms of an "effective stress", regardless of how it is defined. The differential pressure is always needed to evaluate $\bar{C}(P_d)$, while the other term in each expression involves $(P_c - \bar{n}\,P_p)$. Consider, for example, the measurements made by Nur and Byerlee [1971] of the compression of Weber sandstone. Pore strains were measured as a function of the confining pressure, for various values of the pore pressure (Fig. 5.10a). In Fig. 5.10b, the bulk strains are plotted against the effective pressure $P_c - \bar{n}_b P_p$, where $\bar{n}_b$ is calculated from eqn. (5.18). When plotted in this manner, the strains collapse onto a single curve. Although Nur and Byerlee plotted the total bulk strain against the effective stress $(P_c - n_b P_p)$ ($<P>$ in their notation), the differential pressure $(P_c - P_p)$ ($<P'>$ in their notation) was needed to first evaluate n_b. Therefore, while the total strain seems to correlate with $(P_c - n_b P_p)$, the strain cannot in fact be determined without knowledge of *both* P_c and P_p, which is consistent with the conclusions presented above. It is nevertheless true, however, that the integrated stress-strain behavior will be known if $\bar{C}_{bc}$ is known, and C_{bc} can be measured at zero pore pressure. In other words, the elastic behavior of a sandstone can be investigated through experiments carried out at zero pore pressure, if desired. This fact of course allows great simplification of laboratory compressibility tests, although it perhaps has not always been fully appreciated.

The integrated stress-strain relationships can also be developed within the context of the Carroll-Katsube [1983] approach. First, consider eqn. (4.12) for the increment of the mineral-phase strain. Since C_r is constant, and the porosity is also constant (to first order), eqn. (4.12) can be integrated to yield

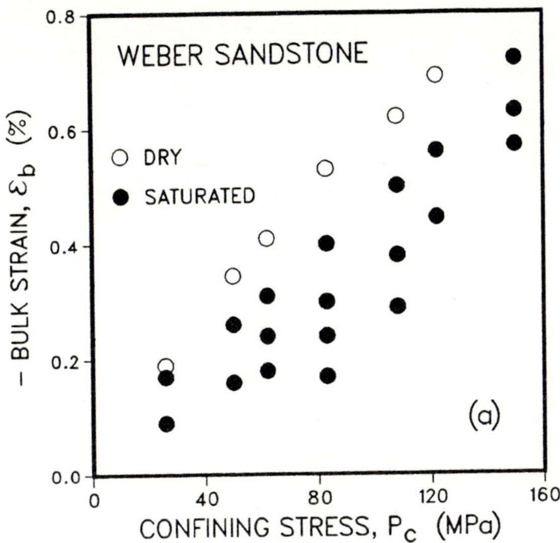

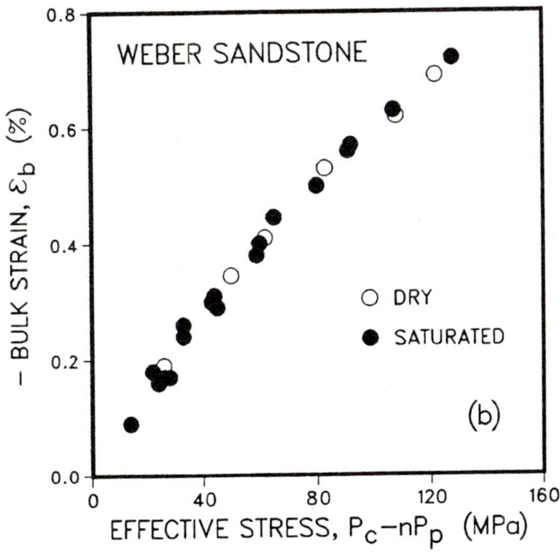

Fig. 5.10. Bulk strain measurements of a Weber sandstone [after Nur and Byerlee, 1971]. In (a) the strains are plotted against P_c, while in (b) they are plotted against $(P_c - \bar{n}_b P_p)$, where $\bar{n}_b$ is calculated in accordance with eqn. (5.18).

$$\varepsilon_r = \frac{-C_r(P_c - \phi P_p)}{1-\phi} . \tag{5.21}$$

Note that equal increments of the pore and confining pressures will not cancel out, but will in fact lead to a decrease in the volume of the mineral phase.

The change in porosity (see eqn. (4.11)) depends only on the differential pressure $P_d = P_c - P_p$, but will be a nonlinear function of P_d, since C_{bc} varies with pressure. Eqn. (4.11) for the porosity change can be integrated as follows:

$$d\phi = -[C_{bc}(1-\phi^i) - C_r]dP_d ,$$

so

$$\phi = \phi^i + \int_{\phi_i}^{\phi} d\phi$$

$$= \phi^i - \int_0^{P_d} [C_{bc}(1-\phi^i) - C_r]dP$$

$$= \phi^i - (1-\phi^i) \int_0^{P_d} C_{bc}(P)dP + C_r P_d$$

$$= \phi^i - (1-\phi^i)\overline{C}_{bc}(P_d)P_d + C_r P_d$$

$$= \phi^i - [(1-\phi^i)\overline{C}_{bc}(P_d) - C_r]P_d . \tag{5.22}$$

Eqns. (5.21) and (5.22) show that the mineral phase strain depends only on the "Terzaghi" effective stress $P_c - \phi P_p$, while the porosity change depends only on the differential pressure $P_c - P_p$.

As an example of the magnitude of the porosity change that can be expected in oilfield situations, consider the measurements made by Zimmerman *et al.* [1986a] on a Bandera sandstone which had a porosity of 16.5% in its unstressed state. An important problem often encountered is that of relating this porosity, which is relatively easy to measure, to that which the rock would have in its natural state. The bulk compressibility, derived by applying eqn.

$$C_{bc} = 5.68\times10^{-7} + 3.68\times10^{-6}\, e^{-P/1208\,\text{psi}}/\text{psi}\ . \qquad (5.23)$$

The secant compressibility function $\overline{C}_{bc}$ can be found by integrating eqn. (5.23):

$$\overline{C}_{bc} = \frac{1}{P_d}\int_0^{P_d} C_{bc}(P)\,dP = 5.68\times10^{-7} - 4.44\times10^{-3}\left[\frac{e^{-P/1208\,\text{psi}} - 1}{P_d}\right]/\,\text{psi}\ . \qquad (5.24)$$

A rock buried at a depth z is typically subjected to a confining pressure that is due to the "lithostatic" gradient, which merely represents the total weight of the rock and fluid lying above depth z. This lithostatic gradient is therefore equal to $\rho_t gz$, where the total density ρ_t is given by $\rho_t = (1-\phi)\rho_r + \phi\rho_f$. With a rock matrix density of 2.65 gm/cc, which is essentially that of quartz, and a liquid density of 1.00 gm/cc, which is that of water, the effective density of the overburden is

$$\rho_t = (1-\phi)\rho_r + \phi\rho_f$$

$$= (1.00-0.165)(2.65\text{ gm/cc}) + (0.165)(1.00\text{ gm/cc})$$

$$= 2.48\text{ gm/cc} = 2480\,\text{kg/m}^3\ . \qquad (5.25)$$

Hence the lithostatic pressure gradient in such a case would be

$$\frac{dP_c}{dz} = \rho_t g = (2480\text{ kg/m}^3)(9.8\text{ m/s}^2)$$

$$= 2.43\times10^{-2}\text{ MPa/m} = 1.07\,\text{psi/ft}\ . \qquad (5.26)$$

Strictly speaking, this would only represent the vertical normal stress. The two horizontal normal stresses are usually less than the vertical stress, but for the purposes of illustration we will assume that the horizontal stresses are equal to the vertical stress, in which case eqn. (5.26) represents the hydrostatic confining pressure. Measured values of horizontal crustal stresses have been reviewed by Jaeger and Cook [1979, pp. 374-378].

Assuming that all of the pores are interconnected, so that the fluid pressure can be freely transmitted, the natural pore pressure would correspond to the hydrostatic pressure of a column of water of depth z:

$$\frac{dP_p}{dz} = \rho_w g = (1000 \text{ kg/m}^3)(9.8 \text{ m/s}^2)$$

$$= 0.98 \times 10^{-2} \text{ MPa/m} = 0.43 \text{ psi/ft}. \quad (5.27)$$

The *in situ* pore pressure may of course have a value different than this, particularly in reservoirs or aquifers that are overlaid with impermeable strata. In these cases, the pore fluid is not in "communication" with the atmospheric pressure at the surface, and abnormally high pore pressures may occur. A well-known example of this phenomenon is in the Gulf Coast, where pore pressures much higher than $\rho_w gz$ have been observed [Swanson *et al.*, 1986]. But again, for the purposes of illustration, we will assume a hydrostatic gradient of 0.43 psi/ft.

Assuming that this rock was buried at a depth of 3 km (9843 ft), the lithostatic confining pressure would be 10,532 psi, and the hydrostatic pore pressure would be 4232 psi. The porosity measured at 0.165 at zero stress, and the matrix compressibility was estimated to be 1.56×10^{-7}/psi. Eqn. (5.21) then predicts that the *in situ* strain in the rock grains would be

$$\varepsilon_r = \frac{-(1.56 \times 10^{-6}/\text{psi})[10{,}532 - (0.165)(4232)] \text{ psi}}{1.00 - 0.165}$$

$$= -0.00184 = -0.184\%. \quad (5.28)$$

The *in situ* porosity at 3000 m can be calculated by first determining the secant compressibility from eqn. (5.24), and then using eqn. (5.22). At a confining pressure of 10,532 psi and a pore pressure of 4232 psi, the differential pressure would be 6300 psi. The secant compressibility is given by eqn. (5.24) as 1.27×10^{-6}/psi. Eqn. (5.22) then shows the porosity at this stress state to be

$$\phi = 0.165 - [(0.835)(1.27 \times 10^{-6}) - 1.56 \times 10^{-7}](6300)$$

$$= 0.159. \quad (5.29)$$

Although this is not very different in absolute terms from the bench-top (zero-stress) value of 0.165, the relative difference of 3.4% could prove important when estimating oil reserves.

Chapter 6. Undrained Compression

The equations developed and discussed in Chapter 2 for pore and bulk strains are appropriate for processes in which the pore pressure and confining pressure can be varied independently. An example of such a process is the reduction in pore pressure due to the withdrawal of fluid from a petroleum or groundwater reservoir. Since the confining stresses acting on the reservoir rock are in general due to the lithostatic gradient, along with tectonic forces, they will not change while the fluid is being produced. The pore pressure will decrease, while the confining pressure remains constant, so the pore and bulk strains will be equal to $C_{pp}dP_p$ and $C_{bp}dP_p$, respectively. There are other situations, however, in which the pore and confining pressures are coupled to each other, and are not independent. For example, during drilling the *in situ* confining stresses will be altered in the vicinity of the borehole. This will induce changes in the pore volume within the affected region of rock. Initially, this will lead to a change in the pore pressure. If the rock is highly permeable, pore fluid will simply flow in such a manner as to re-establish pore pressure equilibrium with the adjacent regions of rock. For rocks that are not very permeable, such as shales, the time required for pore pressure equilibrium to be re-established would be much larger than the time scale of drilling, so that for all intents and purposes the pore fluid must be considered to be ''trapped'' inside the stressed region near the borehole. This phenomenon is discussed by Black *et al.* [1985], who relate it to the poor drillability properties of shales [see also Warren and Smith, 1985; Peltier and Atkinson, 1987]. In these situations, the pore and bulk strains will still be equal to the expressions in eqns. (1.9) and (1.10), but since these stress increments will no longer be independent, these particular equations will not be convenient. For example, if a confining pressure increment dP_c is rapidly applied to an impermeable rock, it would be incorrect to say that $d\varepsilon_b = -C_{bc}dP_c$, since this would not take into account the effect of the induced pore pressure.

For this type of ''undrained'' compression, the pore fluid is not free to move into or out of the pore space in order to establish pore pressure equilibrium, at least not within the time frame of interest. The pore pressure will no longer be an independent variable, but will be related to the confining pressure and the compressibilities of the pores and the pore fluid. It is usually also assumed that undrained compression is relevant to wave propagation problems, where the viscosity of the pore fluid will not permit it to travel between the pores within the time frame (*i.e.*, one period) of the stress oscillations. Undrained behavior is also of importance in rocks of very low permeability, for which the time needed to re-establish pore pressure equilibrium may be extremely long.

The most important parameter for undrained compression is the partial derivative of the bulk strain with respect to the confining pressure, with the mass of fluid in the pores held constant. This compressibility, which can be denoted by C_u, is analogous to C_{bc}, but in general has a different numerical value. Gassmann [1951a] presented an expression for the undrained compressibility C_u in terms of the porosity of the rock, the drained compressibility of the rock, and the compressibilities of the pore fluid and rock grains. Gassmann's expression can be derived within the context of the results of Chapter 2 by starting with eqns. (1.9) and (1.10) for the two volumetric strains:

$$d\varepsilon_b = C_{bp}\,dP_p - C_{bc}\,dP_c , \tag{6.1}$$

$$d\varepsilon_p = C_{pp}\,dP_p - C_{pc}\,dP_c . \tag{6.2}$$

If dP_c is considered to be the independent variable, then there are three unknowns $\{d\varepsilon_b, d\varepsilon_p, dP_p\}$ in the two equations (6.1) and (6.2). A third equation is found by noting that since the mass of the pore fluid is constant, $dV_f/V_f = -C_f\,dP_f$, where the subscript f refers to the pore fluid. Since the pore fluid completely fills the pore space, $V_f = V_p$ and $P_f = P_p$. This assumes that the pore pressure is uniform throughout a certain region of pore space, which is certainly true for quasi-static processes. Hence a third equation is found, without introducing a further unknown:

$$d\varepsilon_p = dV_p/V_p = dV_f/V_f = -C_f\,dP_f = -C_f\,dP_p ,$$

$$i.e., \;\; d\varepsilon_p = -C_f\,dP_p . \tag{6.3}$$

If the confining pressure is considered to be an adjustable parameter, eqns. (6.1-6.3) can be viewed as three linear equations in the three unknowns $\{d\varepsilon_b, d\varepsilon_p, dP_p\}$. These equations can be written in matrix form as

$$\begin{bmatrix} 1 & 0 & -C_{bp} \\ 0 & 1 & -C_{pp} \\ 0 & 1 & C_f \end{bmatrix} \begin{bmatrix} d\varepsilon_b \\ d\varepsilon_p \\ dP_p \end{bmatrix} = \begin{bmatrix} -C_{bc} \\ -C_{pc} \\ 0 \end{bmatrix} dP_c . \tag{6.4}$$

This set of equations can be solved by Cramer's rule of linear algebra. The ratio of $-d\varepsilon_b$ to dP_c, which is the "undrained compressibility", is found to be related to the previously defined porous rock compressibilities by

$$C_u = -\left[\frac{d\varepsilon_b}{dP_c}\right]_{undrained} = C_{bc} - \frac{C_{bp}C_{pc}}{(C_{pp}+C_f)} \,. \tag{6.5}$$

Since all of the compressibilities that appear in eqn. (6.5) are non-negative, it is clear that C_u must be less than the drained compressibility C_{bc}. This reflects the additional stiffness that the trapped fluid imparts to the overall rock-fluid system. If eqns. (2.6-2.8) are used to express C_{pp}, C_{pc} and C_{bp} in terms of C_{bc}, C_r and ϕ, eqn. (6.5) can also be written as

$$C_u = \frac{\phi C_{bc}(C_f - C_r) + C_r(C_{bc} - C_r)}{\phi(C_f - C_r) + (C_{bc} - C_r)} \,. \tag{6.6}$$

Brown and Korringa [1975] extended this relation to the case where the matrix material is neither homogeneous nor isotropic, and concluded that the more general result differs only slightly in form from Gassmann's expression. This generalization involves an additional compressibility C_ϕ, which is equivalent in the current notation to $C_{pc}-C_{pp}$, but does not necessarily equal C_r:

$$C_u = \frac{\phi C_{bc}(C_f - C_\phi) + C_r(C_{bc} - C_r)}{\phi(C_f - C_\phi) + (C_{bc} - C_r)} \,. \tag{6.7}$$

C_ϕ in eqn. (6.7) is the compressibility that relates pore volume strain to the pressure in an undrained compression. Examination of the Brown-Korringa equation under commonly occurring values of the parameters, however, reveals that C_u is almost completely insensitive to deviations of C_ϕ from C_r. For example, consider the data from the Fort Union sandstone discussed by Murphy [1984]. This sandstone had a porosity of 8.5%, a pore compressibility C_{pp} of 11.8×10^{-4}/MPa, and a matrix compressibility of 0.286×10^{-4}/MPa. The Gassmann equation (6.6) then predicts an undrained compressibility under water-saturated conditions of 0.573×10^{-4}/MPa. If C_ϕ differed from C_r by as much as a factor of two, *i.e.*, 0.572×10^{-4}/MPa or 0.143×10^{-4}/MPa, the Brown-Korringa equation (6.7) would predict values of C_u of 0.561×10^{-4}/MPa and 0.580×10^{-4}/MPa, a change of less than 2%.

An important assumption which is implicit in the derivation of eqn. (6.6) is that it is possible to define a "pore pressure" that is constant over regions large enough to contain many pores. If the pore pressure varies from pore to pore, or within a single pore, the use of eqn.

(6.6) would not be justified, and the problem would no longer be determinate without much additional information. This poses no restrictions in applications to the static behavior of a porous rock mass, but does impose restrictions on the rapidity of the processes to which eqn. (6.6) is applicable. These restrictions involve the viscosity of the pore fluid, and the permeability of the rock. Mavko and Nur [1975] and Cleary [1978] have examined the problem of delineating the different response regimes, which are determined by whether or not the pore pressure has sufficient time to equilibrate itself locally. This is in general a difficult question to answer, for a number of reasons. For one, the relevant "permeability" that applies is probably not the usual macroscopic permeability that is used, for example, in reservoir calculations, but is instead a "local" permeability that relates to microscopic flow between adjacent pores [Berryman, 1988]. This leads, in fact, not to two but to three regimes of behavior, as described by Cleary [1978]. Roughly speaking, however, the time constant needed for the achievement of local equilibrium over a region whose characteristic dimension is L is on the order of

$$\tau = \frac{\phi \mu C_t L^2}{k} , \qquad (6.8)$$

where k is the permeability of the rock, μ is the viscosity of the fluid, and $C_t = C_f + C_{pp}$. A typical value for C_t would be 5×10^{-4}/MPa, which is roughly equal to the compressibility of water. If we consider a typical rock core whose characteristic length is a few centimeters, the pore pressure relaxation time will be roughly a few seconds or minutes for a typical reservoir sand whose permeability is on the order of 100 millidarcies, and will be a few hours or days for a shale whose permeability is on the order of microdarcies. (Recall that 1 darcy is approximately equal to $10^{-12}\,\mathrm{m}^2$.)

There are several limiting cases for which the undrained compressibility can be predicted by simpler considerations. Although some of these cases may not be physically meaningful for fluid-saturated sandstones, they can still serve as useful checks on the validity of eqn. (6.6). For very low porosities, *i.e.*, $\phi \rightarrow 0$, eqn. (6.6) reduces to the obvious result that $C_u = C_r$. If the fluid compressibility exactly equals that of the rock, eqn. (6.6) again predicts that $C_u = C_r$, as expected. As the fluid compressibility becomes infinite, such as for rocks saturated with gas (where C_f / C_r may be as high as $10^4 - 10^6$), C_u will approach C_{bc}, proving that the results of Chapters 2-5 will be applicable to the drained *or* undrained compression of gas-saturated rocks. At the other extreme, that of a truly incompressible fluid, eqn. (6.6) takes the following form:

$$\frac{C_u}{C_r} = \frac{C_{bc}(1-\phi) - C_r}{C_{bc} - (1+\phi) C_r} . \qquad (6.9)$$

Liquids are often considered incompressible, and this is a convenient approximation for many engineering calculations. However, in the context of undrained compressibility of sandstones, C_r forms a natural standard by which to judge fluid compressibility, and the ratio C_f/C_r is always much greater than 1.0 in practical situations [Zimmerman, 1985b]. Further manipulation of eqn. (6.6) shows that C_{pc}/C_f is also an important ratio. This dimensionless parameter measures the tendency of the confining pressure to close the pores, relative to the tendency of the pore fluid to resist that closure. Since C_{pc} is often much greater than C_r, the ratio C_{pc}/C_f is not always small.

The full range of Gassmann's relation (6.6) shown in Fig. 6.1 for a Fort Union sandstone, where the undrained compressibility is plotted against the compressibility of the pore fluid. This rock has a porosity of 8.5%, a drained bulk compressibility C_{bc} equal to 1.31×10^{-4}/MPa, and C_r equal to 0.286×10^{-4}/MPa [Murphy, 1984; Zimmerman, 1985b]. The undrained bulk compressibility approaches asymptotic values as the compressibility of the pore fluid goes to zero or infinity. C_u is most sensitive to C_f when C_f is approximately equal to the pore compressibility C_{pp}, which for the Fort Union sandstone is 1.18×10^{-3}/MPa. If the

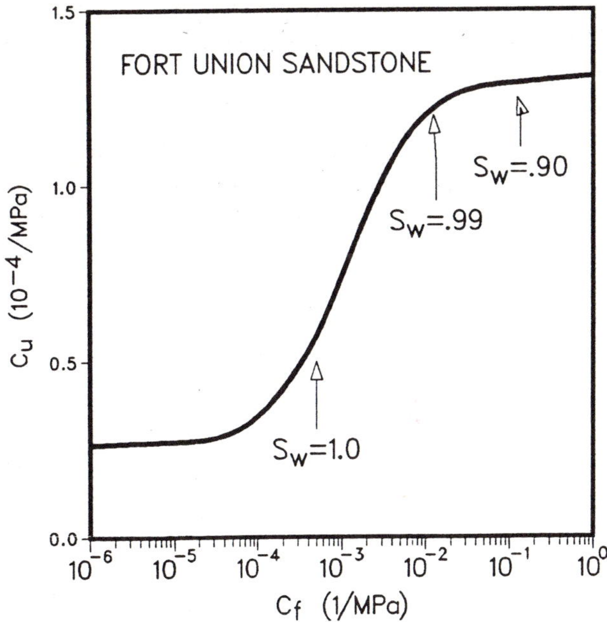

Fig. 6.1. Undrained bulk compressibility of the Fort Union sandstone described by Murphy [1984], as a function of the compressibility of the pore fluid. The water saturation is S_w, and the air saturation is $1-S_w$.

rock were saturated with air, which has a compressibility of 1.25/MPa, the undrained compressibility would essentially (to three digits) equal the drained value, 1.31×10^{-4}/MPa. If the rock were saturated with water, which has a compressibility of 5×10^{-4}/MPa, eqn. (6.6) would yield an undrained compressibility of 0.573×10^{-4}/MPa. An incompressible pore fluid would lead to an undrained compressibility (see eqn. (6.9)) of 0.261×10^{-4}/MPa. The assumption that water is incompressible would therefore yield a grossly incorrect value for C_u. In fact, the water-saturated value of C_u is closer to the air-saturated value than it is to the $C_f \to \infty$ limit.

For quasi-static loadings, the equivalent compressibility of the pore fluid equals the volumetrically-weighted average of the compressibilities of the different fluid phases. Hence any small amount of air tends to greatly increase the effective compressibility of the pore fluid. In Fig. 6.1, the effective compressibilities of various water-air mixtures (at atmospheric pressure) are indicated on the abscissa. As little as 1% air saturation in the pores would be sufficient to lead to a value of C_u very near to the "drained compressibility" limit, where $C_u = C_{bc}$. As in Fig. 6.1, complete liquid saturation typically leads to an undrained compressibility that lies somewhere midway between the two asymptotic values corresponding to an infinitely compressible pore fluid and a truly incompressible pore fluid. Since the compressibility of a gas is inversely proportional to the pressure, the effect of air that is shown in Fig. 6.1 will be greatly minimized at higher pore pressures.

Another phenomenon that is closely related to undrained compression is that of induced pore pressure. If a fluid-saturated porous rock is subjected to undrained compression, the confining pressure would tend to cause the pores to contract, thereby pressurizing the trapped pore fluid. The magnitude of this induced pore pressure increment can be found from eqns. (6.1-6.3) to be

$$dP_p[\text{induced}] = \frac{C_{pp}+C_r}{C_{pp}+C_f}\, dP_c = B\, dP_c\,. \qquad (6.10)$$

The coefficient that relates dP_p to dP_c is usually denoted by B in the field of soil mechanics, where induced pore pressures are of major importance [Mesri *et al.*, 1976]. First, note that since $C_f > C_r$ for practical cases, B will lie between 0 and 1.0. Furthermore, since C_{pp} is typically much larger than C_r,

$$B \approx \frac{C_{pp}}{C_{pp}+C_f} = \frac{1}{1+C_f/C_{pp}}\,. \qquad (6.11)$$

If the sandstone is saturated with a gas, C_f will be much greater than C_{pp}, and B will approach zero. In other words, a gas-saturated sandstone will never develop appreciable induced pore pressures. For liquid saturation, the value of B typically will have a non-negligible value, and will vary with stress. To see this, note that at low effective stresses, a consolidated sandstone will have $C_{pp} \gg C_f$, and so B will approach unity. At higher effective stresses, the ratio C_f / C_{pp} will usually be on the order of unity, so that B will lie somewhere between 0 and 1.0, but not too close to either of these values.

An extensive theoretical and experimental discussion of the phenomenon of induced pore pressures in sandstones has been given by Green and Wang [1986]. In their experiments, external confining pressures were applied to kerosene-saturated sandstone cores. The pore spaces were closed off to the surroundings, except for a pore pressure transducer that was placed flush against the core (see Chapter 13). The tests began with both the pore and confining pressures equal to zero. Fig. 6.2 shows the measured pore pressure as a function of confining pressure, for a Berea sandstone and a Tunnel City sandstone. The Berea was described by Green and Wang as being medium-grained, low-rank Mississippian graywacke with 20% porosity. The grains were well-sorted and subangular, with an average grain-diameter of 155 μm. The mineral composition was 80% quartz, 5% feldspar, 8% clay, and 6% calcite. The Tunnel City was a quartz arenite with 23% porosity. The grains were well-sorted and well-rounded, and consisted of 94% quartz, 5% orthoclase, and 1% lithic fragments, while the cement consisted of quartz and orthoclase overgrowths [Green and Wang, 1986]. If the B

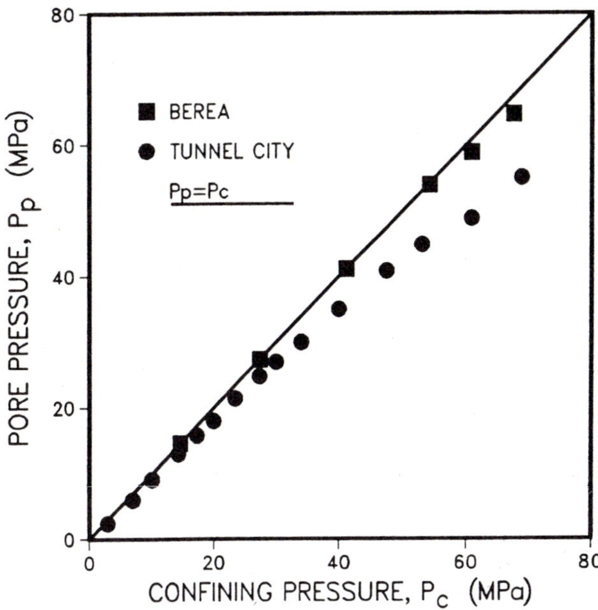

Fig. 6.2. Induced pore pressures in hydrostatically-stressed samples of Berea and Tunnel City sandstone [after Green and Wang, 1986]. The slopes of these data yield the B coefficients.

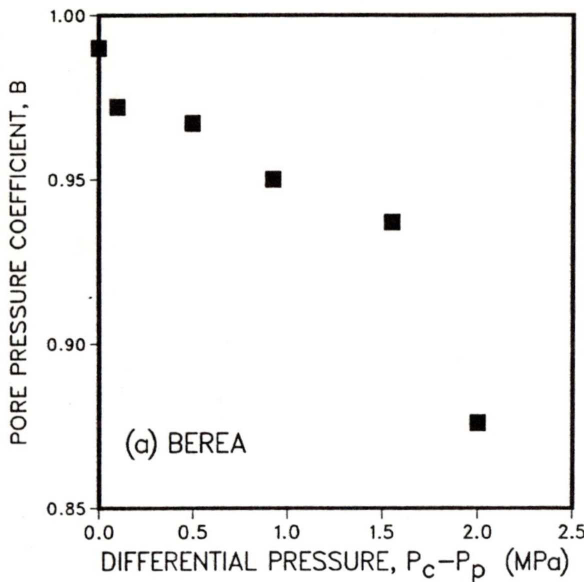

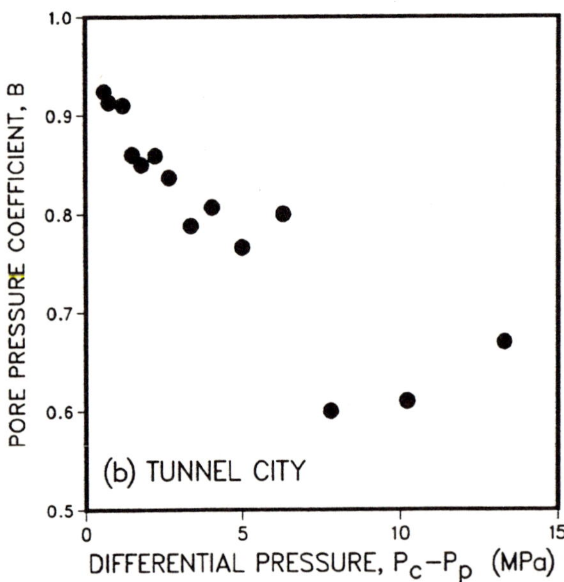

Fig. 6.3. Induced pore pressure coefficients, B, for Berea and Tunnel City sandstones [after Green and Wang, 1986]. Values are found by differentiating the data in Fig. 6.2.

parameter had been equal to unity, the curves would plot as straight lines at a 45° angle. At low pressures, when C_{pp} is high, eqn. (6.10) shows that B will be close to 1.0, and indeed at low pressures the slopes of the curves in Fig. 6.2 are essentially equal to 1.0. As the pressure increases, C_{pp} decreases, and B falls below 1.0. The slopes of these curves, which give the B parameters, are shown in Fig. 6.3 as a function of the differential pressure. Note that if C_{pp} is a function of the differential pressure $P_c - P_p$, then B will be also, since C_f and C_r are essentially independent of pressure. The scatter in the B values merely reflects the inherent scatter in the numerical differentiation of the pressure measurements shown in Fig. 6.2. Green and Wang also compared their measured values of B for the Berea to the values predicted from theory, based on earlier published values of C_{bc} and C_u. They used the expression

$$B = \frac{C_{bc} - C_r}{(C_{bc} - C_r) + \phi(C_f - C_r)} , \tag{6.12}$$

which can be shown to be equivalent to eqn. (6.10). While the predicted value of B decreased from 1.0 to 0.85 as the differential pressure increased from 0 to 2 MPa, the measured value decreased from 1.0 to 0.87. This close agreement corroborates the elastic analysis that led to eqn. (6.10).

Chapter 7. Introduction to Poroelasticity Theory

The discussion of the previous chapters was restricted to processes in which the loads applied to the rock were purely ''hydrostatic''. This term refers to stress states in which the three principal normal stresses are equal in magnitude. The term ''hydrostatic'' is used in reference to the fact that such a state of stress is the only one that can be sustained by a fluid (''hydro'' = water) when it is at rest (''static'' = at rest). However, it should not be assumed that hydrostatic stresses only exist in liquids, or that they are necessarily associated with liquids. While hydrostatic loadings are common in geophysical processes, the more general state of stress is one in which the principal normal stresses are unequal in magnitude. A simple example would be uniaxial stress, in which one principal stress is nonzero, while the other two are equal to zero. In this regard it is worth mentioning a peculiarity of the terminology commonly used in petroleum-related rock mechanics studies. A state of stress in which only one principal stress is non-zero is referred to as ''uniaxial''. The term ''triaxial'' is used to denote a state of stress in which all three principal stresses are non-zero, but two of them are equal in magnitude. The most general case, in which all three principal stresses are non-zero and not necessarily equal, is called ''polyaxial''.

The only stresses that were considered in the earlier chapters were pore pressures or hydrostatic confining pressures. In this chapter, a more general theory of the mechanics of porous rocks will be outlined, which allows for completely arbitrary states of stress. This theory, which was developed chiefly by Biot [1941,1973], is generally referred to as ''poroelasticity''. Although the theory, in its basic form, has existed for several decades, only recently has it been used to solve basic rock mechanics problems such as the state of stress around a borehole [Detournay and Cheng, 1988], and to study phenomena such as the stress-dependence of permeability [Walder and Nur, 1986]. Poroelasticity theory has been extended by Biot [1956a,b] and others [see White, 1983, and references therein] to deal with dynamic processes such as seismic wave propagation. The theory of poroelasticity is fundamentally a higher-order theory than that of classical elasticity, in that it involves concepts such as ''pore pressure'' and ''pore strain'' that have no analogue in classical mechanics [Nunziato and Cowin, 1979]. In this sense, the theory in the previous chapters can be viewed as a special case of poroelasticity for rocks under hydrostatic loading. Poroelasticity not only utilizes concepts and variables that do not exist in classical mechanics, it also predicts certain phenomena which have no counterpart in the classical theory. For example, the equations of dynamic poroelasticity predict the existence of "slow" compressional waves that travel at a speed much less than that of the classical compressional waves. These slow compressional waves attenuate very rapidly as they travel through rock, since they are characterized by motions in which the pore fluid oscillates out of phase with the rock matrix. This solid/fluid counterflow produces large viscous forces at the pore walls, which tends to dissipate energy. Although such waves have been observed in laboratory tests [Plona, 1980], they have yet to be observed in the field.

Another distinction between the theory of poroelasticity and the more "primitive" approach developed in the preceding chapters is that in contrast to the theory of hydrostatic compression, which can be developed in a fully nonlinear form, a usable theory of poroelasticity that accounts for the stress-dependence of the elastic moduli is still in the process of development. Hence the term "poroelasticity" usually refers to the linearized theory of poroelasticity, in which the compressibilities and the other constitutive coefficients are independent of stress. There are many physical processes in which the stress increments are small enough that this restriction is of no consequence, since the governing equations can always be linearized for small stress increments. An example of such a process is wave propagation. Both the naturally occurring waves that result from seismic activity, as well as the artificially created "acoustic" waves used in well logging, typically involve stress increments of only a few megapascals, which are very small in relation to the existing *in situ* stresses. Another process that has been successfully treated by linearized poroelasticity is the oscillation of the water level in wells due to fluctuations in barometric pressure at the surface or to solid-earth tidal forces [van der Kamp and Gale, 1983]. In these problems, the bulk strain increments are on the order of only 10^{-8}, and so the linearized theory is appropriate. Such water-level fluctuations can be used in the inverse problem of estimating the formation compressibility and other poroelastic constitutive parameters. For other problems, such as determining the state of stress around a borehole, linearized poroelasticity may be only a rough first approximation, since the induced stresses will be of the same order of magnitude as the initial stresses, and hence large enough for the nonlinearity of the stress-strain relationship to become an important factor. Some progress has been made in treating the borehole stress problem for a rock with a nonlinear stress-strain relationship [Biot, 1974; Santarelli *et al.*, 1986], but these treatments have not included the effects of pore pressure.

Before presenting the theory of poroelasticity, a brief discussion of the theory of stress is needed. The treatment here will be restricted to the bare minimum that is necessary in order to be able to present the basic equations of poroelasticity. A discussion of other aspects of the theory of stress can be found in elasticity texts such as that by Sokolnikoff [1956]. Consider first a piece of solid material that is subjected to an arbitrary type of loading along its boundaries. At this point in the discussion, the porous nature of sandstone will be neglected. This loading will set up a certain state of stress within the body, in accordance with the equations of equilibrium. (The theory of stress actually applies equally well to solids or fluids, and is certainly not restricted to elastic solids. However, the concepts are probably simpler to learn by considering the material to be an elastic solid.) Now imagine that an infinitesimally small cube is (mentally) ''cut out'' of the body, such as when a ''free-body'' diagram is drawn in elementary mechanics (Fig. 7.1). This cube must be small in order that the stresses within it can be assumed to be essentially constant from point to point. The forces acting on this cube as a result of its interaction with the material external to it can be represented by force vectors **F** that act on each face of the cube. If each of these vectors is ''divided'' by the area of the face upon which the force is acting, the result is a traction vector $\mathbf{t}=(1/A)\mathbf{F}$, which has dimensions of [force/area], *i.e.*, [pressure].

Since the resultant force on any face of the cube will in general vary from face to face, the traction can be thought of as a function of the outward unit normal vector of the surface, *i.e.*, $\mathbf{t}=f(\mathbf{n})$. According to Cauchy's theorem [Sokolnikoff, 1956, pp. 41-42] this function $f(\mathbf{n})$ is a linear transformation, and can therefore be represented by matrix multiplication. The 3×3 matrix that relates **t** to **n** is known as the stress matrix, **T**, *i.e.*,

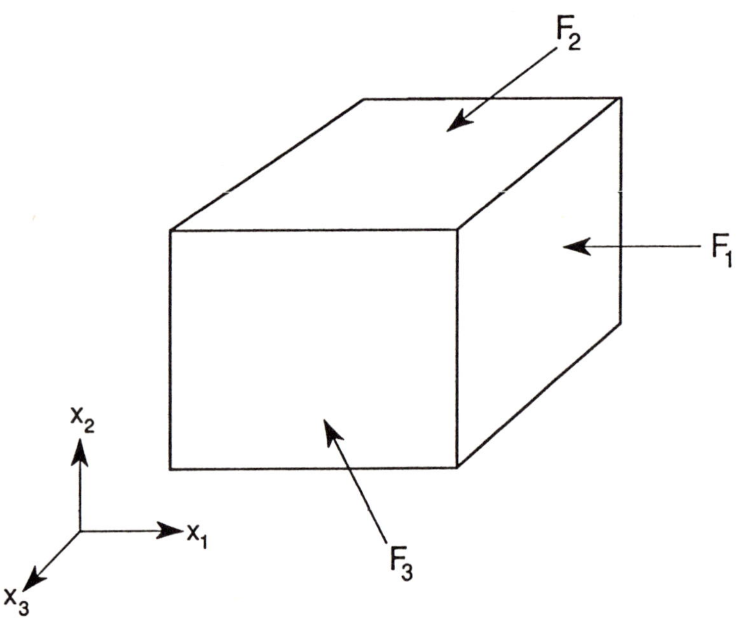

Fig. 7.1. Infinitesimally small cube of rock material, ''cut out'' of a larger piece of rock that is subjected to various forces. The net force transmitted to the cube by the adjacent material can be represented by vectors F acting at the center of each face.

$$\mathbf{t} = \mathbf{T}\,\mathbf{n}\ . \tag{7.1}$$

Since **n** is the *outward* unit normal vector, tractions will be positive if they are *tensile*. The components of **T** have a very clear physical interpretation, which can be seen by considering, for example, the face of the cube whose outward unit normal vector points in the positive x_1 direction (Fig. 7.2). For this face, $\mathbf{n}=(n_1,n_2,n_3) = (1,0,0)$, so eqn. (7.1) yields

$$\mathbf{t} = \begin{bmatrix} t_1 \\ t_2 \\ t_3 \end{bmatrix} = \begin{bmatrix} T_{11} & T_{12} & T_{13} \\ T_{21} & T_{22} & T_{23} \\ T_{31} & T_{32} & T_{33} \end{bmatrix} \begin{bmatrix} 1 \\ 0 \\ 0 \end{bmatrix} = \begin{bmatrix} T_{11} \\ T_{21} \\ T_{31} \end{bmatrix}\ . \tag{7.2}$$

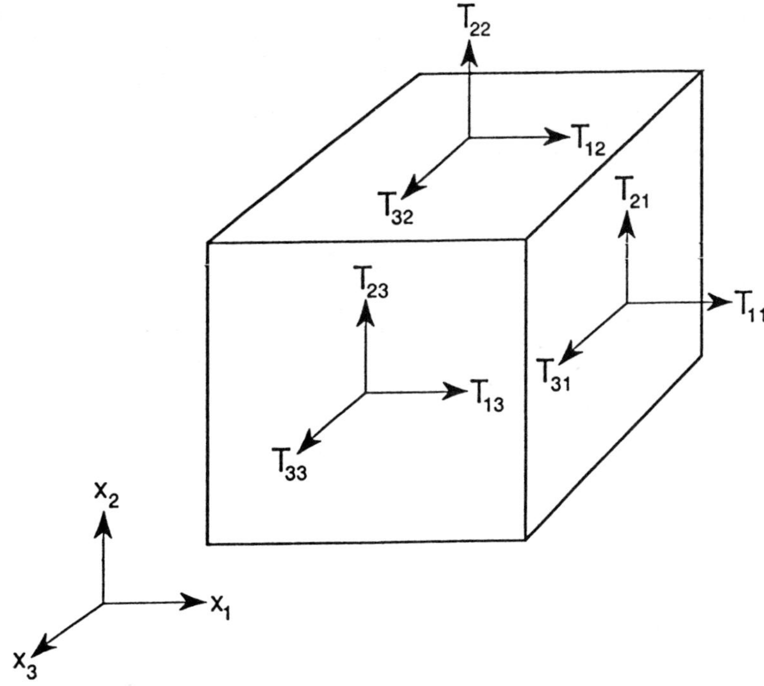

Fig. 7.2. Schematic diagram showing the components of the stress matrix. The force $\mathbf{F}_1$ (see Fig. 7.1) is related to the traction by $\mathbf{F} = A\,\mathbf{t}$, where A is the area of the face of the cube. The traction vector is then resolved into the components $\{T_{11}, T_{21}, T_{31}\}$.

Hence T_{11} represents the normal component of the traction on the face which is perpendicular to the x_1-axis, while T_{21} and T_{31} represent shear components on that face in the x_2 and x_3 directions, respectively. In the commonly used terminology, the distinction between the tractions and the stresses is ignored, and these traction components are referred to as the normal and shear *stresses*, respectively. Although the law of conservation of angular momentum requires that the stress matrix be symmetric, *i.e.*, $T_{12}=T_{21}$, etc., it is convenient to use different symbols for the off-diagonal terms, since they represent physically different tractions which happen to have the same magnitude.

The traction components on the three faces shown in Fig. 7.2 can therefore be read directly from the stress matrix, the columns of which are merely the traction vectors for the three faces perpendicular to the three coordinate axes. However, there is no reason not to be interested in faces which are at oblique angles to the coordinate axes. For example, failure criteria for rocks often require that the maximum shear stress reach some critical value known as the shear strength [Jaeger and Cook, 1979, p. 90]. In order to test whether or not a given state of stress will cause failure, it is obviously necessary to consider the stresses on all possible planes. This can still be done with eqn. (7.1); the resultant normal and shear stresses on an arbitrary plane will be then equal to some combination of the components of the stress matrix **T** and the outward unit normal vector **n**. On certain particular planes, the shear stresses vanish, and the stress state is purely one of a normal traction. Since the stress matrix is always symmetric, the theory of linear algebra can be used to show that the normal stresses on these planes are locally extreme values. Furthermore, this occurs on three planes which are always perpendicular to each other. The outward unit normal vectors to these three planes are known as the *principal stress directions*, and the normal stresses acting on these planes are known as the *principal normal stresses*.

For the purposes of developing the constitutive equations of linear poroelasticity, it is convenient to decompose the stress matrix into two parts, an "isotropic" part $\mathbf{T}^{\mathrm{I}}$ and a "deviatoric" part $\mathbf{T}^{\mathrm{D}}$ [van der Knaap, 1959]. The isotropic part is defined by

$$\mathbf{T}^{\mathrm{I}} = -P_c \mathbf{I} \, , \tag{7.3}$$

where P_c is related to the components of the stress matrix by

$$P_c = -(T_{11}+T_{22}+T_{33})/3 \, . \tag{7.4}$$

The minus sign is needed in eqn. (7.4) to account for the fact that, by convention, tensile normal stresses are considered positive, whereas pressures are considered positive is they are *compressive*. The matrix **I** in eqn. (7.3) is the 3×3 identity matrix, whose diagonal components are all equal to 1, and whose off-diagonal components are all equal to 0. This definition of P_c

reduces to the usual concept of pressure for the special case of a purely hydrostatic stress state, and also plays the role of the pressure in thermodynamic developments of continuum mechanics. The subscript c is used to express the equivalence of this pressure with the "confining pressure" discussed in previous chapters. In the terminology of matrix algebra, $P_c = -(1/3)trace(\mathbf{T})$, where the trace is the sum of the diagonal components. Since the diagonal terms are the normal stresses, $-P_c$ is the *mean* normal stress. The deviatoric part of the stress is then defined such that $\mathbf{T} = \mathbf{T}^{\mathrm{I}} + \mathbf{T}^{\mathrm{D}}$, *i.e.*,

$$\mathbf{T}^{\mathrm{D}} = \mathbf{T} - \mathbf{T}^{\mathrm{I}},$$

so

$$\mathbf{T}^{\mathrm{D}} = \begin{bmatrix} T_{11}+P_c & T_{12} & T_{13} \\ T_{21} & T_{22}+P_c & T_{23} \\ T_{31} & T_{32} & T_{33}+P_c \end{bmatrix} . \tag{7.5}$$

For purely hydrostatic stress states, $\mathbf{T}^{\mathrm{I}} = \mathbf{T}$, and $\mathbf{T}^{\mathrm{D}} = 0$.

In the previous chapters, the only measure of deformation that was used was volumetric strain. This measure of deformation does not fully describe the possible types of deformation that may occur in a rock. For example, since a rock may be compressed by different amounts in the three different directions of a rectangular coordinate system, it is obvious that more than one quantity will be needed to specify the deformation. In general, it has been found that the strain can be fully characterized by a 3×3 strain matrix. In the full generality of *finite* deformation, the strain matrix is related to the displacement vector in a nonlinear manner, and it is consequently difficult to interpret the physical meaning of its components. For some geological problems, the use of nonlinear strain measures is unavoidable, since the deformations that occur over geological time may be quite large [see Jaeger and Cook, 1979, pp. 447-459]. For most engineering and geophysical problems, however, it is acceptable, and very convenient, to use the linearized strain matrix that corresponds to *infinitesimal strain.*

To develop the strain matrix, consider first a solid body that undergoes an arbitrary deformation, as in Fig. 7.3. The displacement of a particle that is initially located at point (x_1, x_2, x_3) is denoted by the vector $\mathbf{u}(x_1, x_2, x_3)$, whose three components are (u_1, u_2, u_3). The components of the strain matrix $\boldsymbol{\varepsilon}$ can then be defined in terms of the derivatives of $\mathbf{u}$ as follows:

$$\varepsilon_{ij} = \frac{1}{2}\left[\frac{\partial u_i}{\partial x_j} + \frac{\partial u_j}{\partial x_i}\right] . \tag{7.6}$$

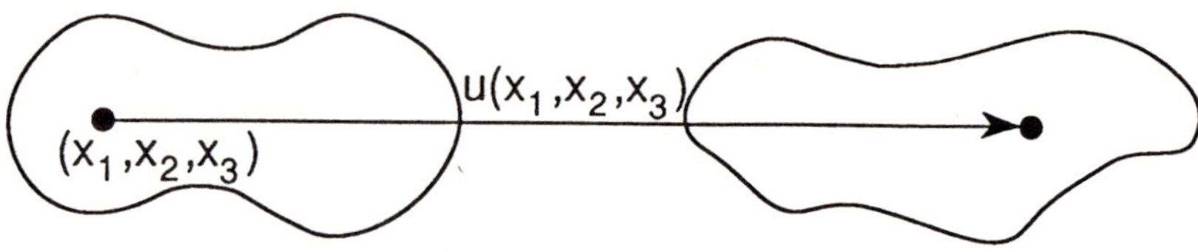

Fig. 7.3. Arbitrary deformation of a solid body. The displacement at each point (x_1,x_2,x_3) is denoted by $\mathbf{u}(x_1,x_2,x_3)$.

The subscripts i and j in eqn. (7.6) are dummy indices that may take on any of the three values 1, 2, or 3. For example, the strain component ε_{12} is defined by

$$\varepsilon_{12} = \frac{1}{2}\left[\frac{\partial u_1}{\partial x_2} + \frac{\partial u_2}{\partial x_1}\right] . \tag{7.7}$$

If the deformation is not small, nonlinear terms involving products of the terms $\partial u_i/\partial x_j$ would also appear in the definitions of the strain components. Since the strain is defined in terms of the derivatives of the displacement, a displacement that does not vary from point to point gives rise to no strain.

The diagonal components of the strain matrix represent fractional elongations in the direction of one of the coordinate axes. This is most easily seen by considering a bar that is stretched uniaxially, as in Fig. 7.4. Consider two points on the bar that are initially located at $x = x_o$ and $x = x_o + dx$. These points are then displaced by amounts $u_x(x_o)$ and $u_x(x_o + dx)$. The initial length of the segment between the points is

$$L^i = (x_o + dx) - (x_o) = dx .$$

The distance between the points after deformation is

$$L = [(x_o + dx + u_x(x_o + dx)] - [x_o + u_x(x_o)]$$

$$= dx + u_x(x_o + dx) - u_x(x_o) .$$

The change in length is given by

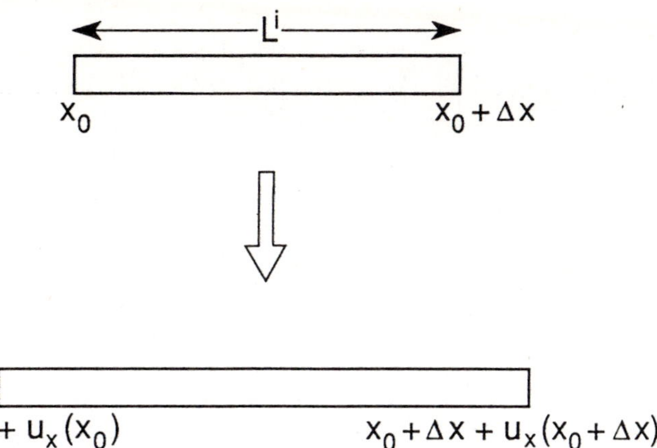

Fig. 7.4. Uniaxial deformation of a thin bar, used in the derivation of eqn. (7.8).

$$\Delta L = L - L^i = [dx + u_x(x_o + dx) - u_x(x_o)] - [dx]$$

$$= u_x(x_o + dx) - u_x(x_o),$$

so the fractional change in length is

$$\frac{\Delta L}{L^i} = \frac{u_x(x_o + dx) - u_x(x_o)}{dx} \rightarrow \left.\frac{\partial u_x}{\partial x}\right|_{x_o} = \varepsilon_{xx}(x_o)\ . \tag{7.8}$$

A similar but somewhat more lengthy analysis shows that the shear strains are a measure of angular distortion [Jaeger and Cook, 1979, pp. 37-39].

When a rock is stressed, there are two qualitatively different modes of deformation that may result. One mode of deformation is pure dilatation, in which the volume changes, but the rock is otherwise undistorted (Fig. 7.5). For example, if three points in the rock are aligned such that they form a right angle before deformation, they will continue to form a right angle after undergoing a pure dilatation. The other type of deformation is one in which the volume remains the same, but the rock distorts (Fig. 7.6). In this sort of deformation, angles will be changed, but the overall volume will be preserved. These two modes correspond to the isotropic and deviatoric parts of the strain matrix:

$$\boldsymbol{\varepsilon}^{\mathrm{I}} = trace\,(\boldsymbol{\varepsilon})\mathbf{I}\ , \tag{7.9}$$

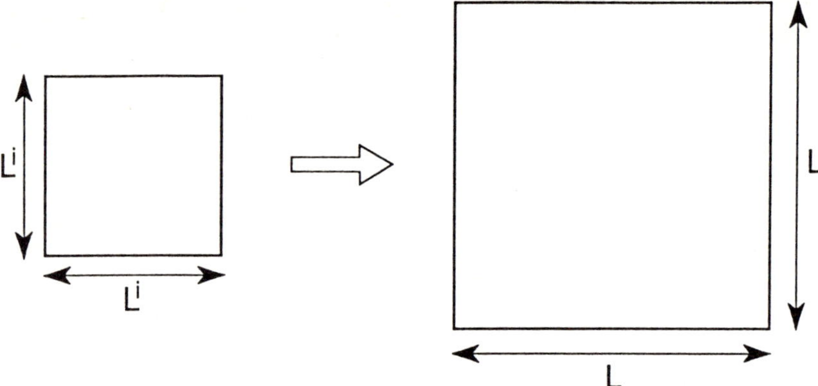

Fig. 7.5. Schematic diagram of a purely dilatational deformation. The square on the left is stretched by equal amounts in all directions; its area changes, but its "shape" remains the same.

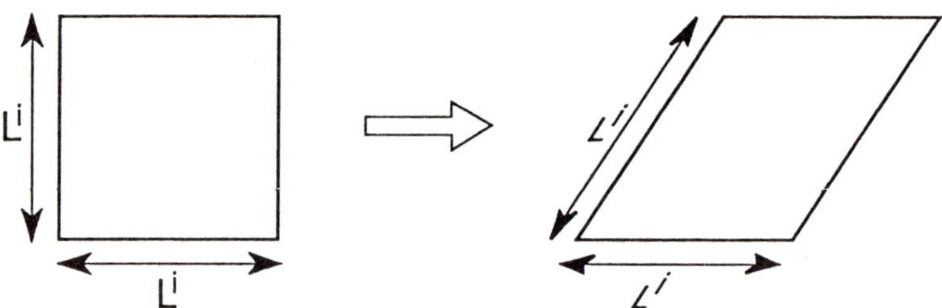

Fig. 7.6. Schematic diagram of a purely deviatoric deformation. The square on the left is distorted into a rhombus, but its area remains unchanged.

$$\boldsymbol{\varepsilon}^{\mathrm{D}} = \boldsymbol{\varepsilon} - \boldsymbol{\varepsilon}^{\mathrm{I}} , \tag{7.10}$$

where, as before, *trace* $(\boldsymbol{\varepsilon}) = \varepsilon_{11} + \varepsilon_{22} + \varepsilon_{33}$. The trace of the strain matrix is also equal to the bulk volumetric strain ε_b. This can be seen by recognizing that if a cube of rock (as in Fig. 7.5) has sides of length L^i before deformation, the lengths of the three sides after deformation will be $L^i(1+\varepsilon_{11})$, $L^i(1+\varepsilon_{22})$, and $L^i(1+\varepsilon_{33})$, respectively. The deformed volume of this cube will be

$$V_b = (L^i)^3(1+\varepsilon_{11})(1+\varepsilon_{22})(1+\varepsilon_{33})$$

$$= V_b^i \left[1 + (\varepsilon_{11} + \varepsilon_{22} + \varepsilon_{33}) + \text{terms of order } (\varepsilon^2) \right] .$$

So, to within first-order in ε,

$$\Delta V_b / V_b^i = \varepsilon_b = (\varepsilon_{11} + \varepsilon_{22} + \varepsilon_{33})$$

$$= \mathit{trace}\,(\boldsymbol{\varepsilon}) . \qquad (7.11)$$

If the stress increments are sufficiently small, it is reasonable to assume that the resulting strains will be linear functions of the stresses. It is at this point that the decomposition of the stresses and strains reveals its usefulness. Within the theory of poroelasticity, the constitutive variables will include the stress matrix and the bulk strain matrix, as well as the pore pressure and the pore strain. These stresses and strains are coupled in the sense that the bulk strain is affected by the pore pressure, etc. However, a partial decoupling is brought about through the isotropic/deviatoric decomposition. First, note that in classical elasticity, the isotropic part of the strain is related only to the isotropic part of the stress, while the deviatoric part of the strain is related only to the deviatoric part of the stress [Billington and Tate, 1981, p. 134]. This is equivalent to saying that shear stresses do not alter the volume, and hydrostatic stresses do not lead to distortion. (This assertion applies only to macroscopically isotropic materials, of course. If the material is anisotropic, equal stresses in different directions will give rise to unequal strains, and so an isotropic stress state may lead to distortion.) When extending these concepts to poroelasticity, it is usually further assumed [Biot, 1941] that pore pressures cause no distortion, and that shear stresses do not alter the pore volume. If the pores are isotropically oriented, which they must be for a macroscopically isotropic rock, this first assumption is obvious. The second assumption follows from the fact that since it is not reasonable for a positive shear stress to increase the pore volume, say, and a negative shear stress to decrease the pore volume, the linear coefficient that relates each shear stress to the pore strain must be zero. Since there is only one independent component of an isotropic matrix, it is convenient to use ε_b and P_c as variables instead of $\boldsymbol{\varepsilon}^{\mathrm{I}}$ and $\mathbf{T}^{\mathrm{I}}$. Hence ε_b will be a linear function of P_c and P_p, $\boldsymbol{\varepsilon}^{\mathrm{D}}$ will be a linear function of $\mathbf{T}^{\mathrm{D}}$, and ε_p will be a linear function of P_c and P_p. Seen from this point of view, the constitutive equations of linear poroelasticity should be an obvious combination of the porous rock compressibility relations (1.9) and (1.10) and the stress-strain equations of classical elasticity:

$$\varepsilon_b = -C_{bc} P_c + (C_{bc} - C_r) P_p , \qquad (7.12)$$

$$\varepsilon_p = -\frac{(C_{bc} - C_r)}{\phi} P_c + \frac{[C_{bc} - (1+\phi)C_r]}{\phi} P_p \, , \tag{7.13}$$

$$\boldsymbol{\varepsilon}^{\mathrm{D}} = \frac{1}{2G} \mathbf{T}^{\mathrm{D}} \, . \tag{7.14}$$

Eqn. (7.14) states that the deviatoric strain tensor is a scalar multiple of the deviatoric stress tensor, and that each component of the deviatoric strain is related to the corresponding component of the deviatoric stress by $\varepsilon_{ij}^{\mathrm{D}} = (1/2G)T_{ij}^{\mathrm{D}}$.

In eqns. (7.12) and (7.13), the relationships between the various porous rock compressibilities that were derived in Chapter 2 have been used. The only additional relationship that appears in the more general poroelasticity equations is the usual relationship between shear stresses from classical elasticity, eqn. (7.14), which involves the shear modulus G. The hybrid notation of eqns. (7.12-7.14), involving both compliances (*i.e.*, the compressibilities) and stiffnesses (*i.e.*, the shear modulus), is used here in order to preserve as much continuity as possible with the notation used in the previous chapters. There are numerous other notations used in poroelasticity, as well as many slight differences in the formulations of the equations that are more than merely notational. Commonly used formulations are presented by Rice and Cleary [1976], and Green and Wang [1986].

The equations that relate the stresses to the strains, which are known in mechanics as the "constitutive relations", are not sufficient in themselves to allow the solution of boundary value problems. To these equations must be appended the equations of stress equilibrium, which express the principle of conservation of momentum. In the absence of body forces or inertia terms, the equations of stress equilibrium can be expressed as

$$div\,\mathbf{T} = \mathbf{0} \, , \tag{7.15}$$

where *div* represents the divergence operator. In rectangular coordinates, eq. (7.15) takes the form of three scalar equations that can be written as

$$\frac{\partial T_{11}}{\partial x_1} + \frac{\partial T_{12}}{\partial x_2} + \frac{\partial T_{13}}{\partial x_3} = 0 \, ,$$

$$\frac{\partial T_{21}}{\partial x_1} + \frac{\partial T_{22}}{\partial x_2} + \frac{\partial T_{23}}{\partial x_3} = 0 \, ,$$

$$\frac{\partial T_{31}}{\partial x_1} + \frac{\partial T_{32}}{\partial x_2} + \frac{\partial T_{33}}{\partial x_3} = 0 \, . \tag{7.16}$$

The additional equations that are needed depend on the type of motion that is being described. For purely static problems, or steady-state conditions, the only additional equation that is needed is one for the pore pressure field. The pore pressure is often found from the equations of fluid statics,

$$\frac{dP_p}{dz} = -\rho_f g \ , \tag{7.17}$$

where ρ_f is the fluid density and z is the vertical coordinate, along with boundary conditions (such as atmospheric fluid pressure at the surface $z = 0$). This static regime corresponds to what was referred to as "drained" behavior in Chapter 6.

In the next regime of behavior, which can be called quasi-static, the deformations are slow enough that inertial effects can be ignored, but the changes in pore pressure that result from pore fluid diffusion must be accounted for. The diffusion of pore pressure can be treated by the usual equation of reservoir engineering,

$$\frac{\partial P_p}{\partial t} = \frac{k}{\phi \mu C_t} \nabla^2 P_p \ , \tag{7.18}$$

in which k is the formation permeability, μ is the pore fluid viscosity, and C_t is the total compressibility of the pore space plus the pore fluid [Matthews and Russell, 1967, p. 7]. This coupling between the rock stresses and the pore pressure leads to interesting phenomena, whose importance to petroleum engineering has only recently begun to be examined. Detournay and Cheng [1988], for instance, showed that, because of transient changes in pore pressure, the borehole stresses in the first few hours after drilling may be appreciably, and qualitatively, different than those predicted by the Kirsch solution of classical elasticity [Sokolnikoff, 1956]. Rice and Cleary [1976] solved some basic problems in the quasi-static theory, such as finding the stresses arising from fault-like motions along planar discontinuities. Palciauskas and Domenico [1989] used quasi-static poroelasticity to study the long-term compaction of sedimentary basins. Walder and Nur [1986] used poroelasticity theory to infer the existence of a sample-size dependence during measurements of permeability by the pulse-decay method.

PART TWO: COMPRESSIBILITY AND PORE STRUCTURE

Chapter 8. Tubular Pores

In Part One, the relationships between the various compressibilities were discussed from a theoretical point of view, and experimental data were presented to verify these relationships. Lower bounds were derived for the compressibilities, although it was seen that compressibilities often greatly exceed the lower bounds. At the same time, there are no finite upper bounds to constrain the compressibility values. Sandstones exhibit wide variation in their compressibility, both from one rock to another, but also as a function of pressure for a given rock. The factors which determine the numerical value of the compressibility, and which therefore distinguish the mechanical behavior of different rocks, are mineral composition and pore structure. It was mentioned in Chapter 2 that the mineral composition *per se* has a relatively minor influence, and can be accounted for with adequate accuracy by using the Voigt-Reuss-Hill average for the effective elastic moduli of the mineral phase. This requires only knowledge of the volume fractions and elastic moduli of the various minerals. This latter information is known for most rock-forming minerals [Simmons and Wang, 1971; Clark, 1966], although compressibility data for clays are somewhat scarce [Wang *et al.*, 1980; Tosaya and Nur, 1982], due to the difficulty of obtaining samples of ''pure clay'' that are large enough to perform tests on.

The main factor that determines the compressibility, and the factor which controls the precise variation of compressibility with stress, is the pore structure. Due to the irregularity and heterogeneity of the pore structure of sandstones, modeling their mechanical behavior is no simple matter. The following chapters will discuss various models that have been proposed, and methods that have been used, to quantify the relationship between pore structure and compressibility.

In principle, if the precise pore geometry of a particular sandstone were known, the compressibility could be calculated by solving the equations of elasticity for the rock, with the appropriate boundary conditions over the exterior and interior (pore) surfaces. In practice, however, this is never feasible. Part of the difficulty is that, in three dimensions, the elasticity equations can only be solved analytically for pores which have geometrically regular shapes such as spheres, ellipsoids, etc., which do not entirely represent the true shapes of pores in sandstones. Furthermore, with the exception of some extremely simplified cases such as regular arrays of spherical pores, analytical solutions for bodies with many pores cannot be obtained.

Although numerical methods such as finite elements [Gudehus, 1977] and boundary elements [Venturini, 1983] can treat irregular geometries, it is as yet not computationally feasible to simulate the compression of a sandstone with a realistic pore structure. Progress in the area of relating compressibility to pore structure has thus far centered around finding analytical solutions for isolated pores of certain idealized shapes, and then using some approximate method to account for the fact that rocks contain numerous pores whose stress and strain fields interact.

One of the earliest models of a pore to be used in attempts to correlate mechanical properties of rocks to their pore structures was that of a cylindrical tube of circular or elliptical cross-section. This model has been used to study the attenuation of elastic waves in fluid-filled porous media [Biot, 1956a,b], and to study the effect of porosity and grain diameter on permeability [Scheidegger, 1974], for example. Scanning electron micrographs of sandstones [Weinbrandt and Fatt, 1969; Pittmann, 1984] show that the cross-sections of real pores are never as simple or as smooth as circles or ellipses. Actually, a somewhat irregular polygonal shape is a more accurate model of the pore bodies in sandstones, at least in two-dimensional cross-section. Nevertheless, the elliptical pore is an appropriate starting point for attempts to relate the numerical value of the compressibility to the pore structure.

The compressibility of a tubular pore of elliptical cross-section under conditions of plane-stress has been found by Walsh *et al.* [1965]. This was done by solving the appropriate boundary-value problem in elasticity, with the assumption of zero stresses on the surface of the cavity, and hydrostatic stresses of magnitude P at infinity [Inglis, 1913; Pollard, 1973; Bernabe *et al.*, 1982; Fig. 8.1]. The change in the cross-sectional area of the hole was found by integrating the displacement around the perimeter of the hole, and the pore compressibility C_{pc} is then found by dividing the area change by the applied pressure and the initial pore area.

In elliptical coordinates, the normal displacement at the surface of an elliptical hole in a hydrostatically stressed body, whose semi-major and semi-minor axes are a and b, is given by

$$u_n = \frac{P}{hE}\left[(a^2+b^2) - (a^2-b^2)\cos 2\beta\right], \tag{8.1}$$

where E is the Young's modulus of the body, h is the metric coefficient of the elliptical coordinate system, and β is the "angular" coordinate. (The value of the metric coefficient is not relevant to the present calculation, since it eventually cancels out of the required integration.) The change in the cross-sectional area of the hole is found by integrating the normal displacement over the surface of the hole:

$$\Delta A = \int_0^{2\pi} u_n\,(h\,d\beta)$$

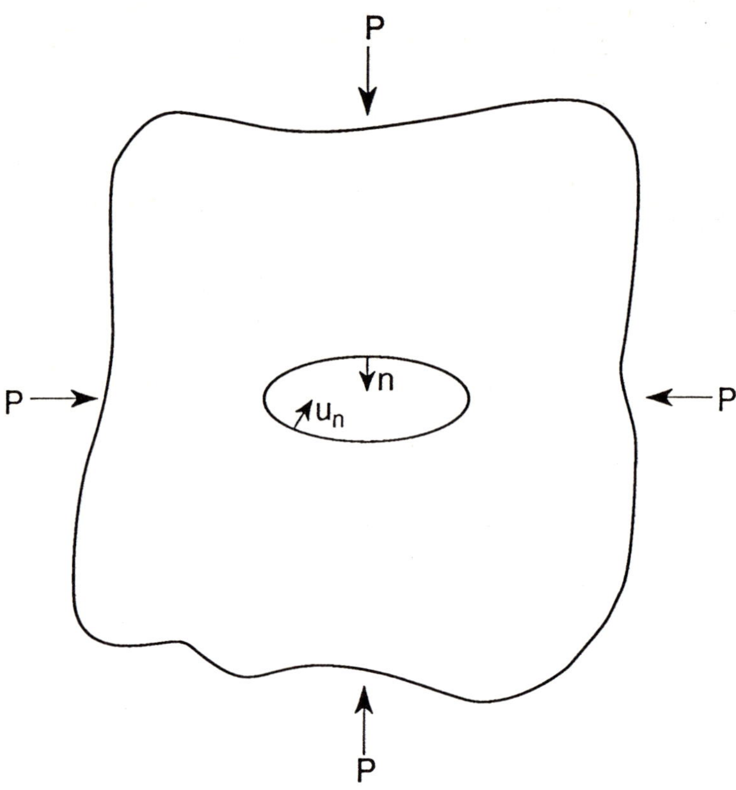

Fig. 8.1. Idealized problem in elasticity that must be solved in order to find the pore compressibility of an isolated tubular pore of elliptical cross-section.

$$= \int_0^{2\pi} \frac{P}{E} \left[(a^2+b^2) - (a^2-b^2)\cos 2\beta \right] d\beta$$

$$= \frac{2\pi (a^2+b^2) P}{E} \,. \tag{8.2}$$

Since the initial area of the hole is πab, the plane-stress pore compressibility is

$$C_{pc} = \frac{\Delta A}{A^i P} = \frac{2\pi(a^2+b^2)P}{\pi abEP} = \frac{2}{E}\left[\frac{a}{b} + \frac{b}{a}\right]$$

$$= \frac{2}{E}\left[\alpha + \frac{1}{\alpha}\right], \tag{8.3}$$

where $\alpha = b/a \leq 1$ is the aspect ratio of the hole.

The expression derived by Walsh *et al.* [1965] for C_{pc}, which is appropriate for plane-stress problems, must be modified somewhat to render it more directly applicable to tubular pores in sandstones. Plane-stress is a state in which the stresses are zero on all planes which are perpendicular to the axis of the tube. This approximation, although it is never completely accurate [Sokolnikoff, 1956], is nevertheless useful for the in-plane loading of thin plates. A more appropriate idealization for a long-tubular pore in a rock, however, is that of plane-strain, in which the strains are zero in the direction parallel to the axis of the tube. Plane-stress solutions can be transformed into plane-strain solutions by first expressing the results in terms of the shear modulus G and Poisson's ratio ν, and then replacing ν with $\nu/(1-\nu)$. Since $E = 2G(1+\nu)$, the plain-strain version of eqn. (8.3) is

$$C_{pc} = \frac{(1-\nu)}{G}\left[\alpha + \frac{1}{\alpha}\right]. \tag{8.4}$$

The plain-strain pore compressibility of the elliptical hole is plotted in Fig. 8.2, in both log-log and semilog format. An interesting feature of expression (8.4) is that C_{pc} depends on the elastic moduli of the rock material only through the combination $(1-\nu)/G$, which is one of the "canonical" moduli introduced by Dundurs [1967]. Circular pores correspond to the limiting case $\alpha = 1$, for which the pore compressibility reduces to

$$C_{pc} = \frac{2(1-\nu)}{G}. \tag{8.5}$$

For small deviations from circularity, eqn. (8.4) can be expanded in terms of the eccentricity parameter $e = (1-\alpha)$ to yield

$$C_{pc} = \frac{(1-\nu)}{G}\left[(1-e) + \frac{1}{1-e}\right]$$

$$= \frac{(1-\nu)}{G}\left[(1-e) + (1+e+e^2+\cdots)\right]$$

$$= \frac{2(1-\nu)}{G}\left[1+\frac{e^2}{2}+\cdots\right]. \tag{8.6}$$

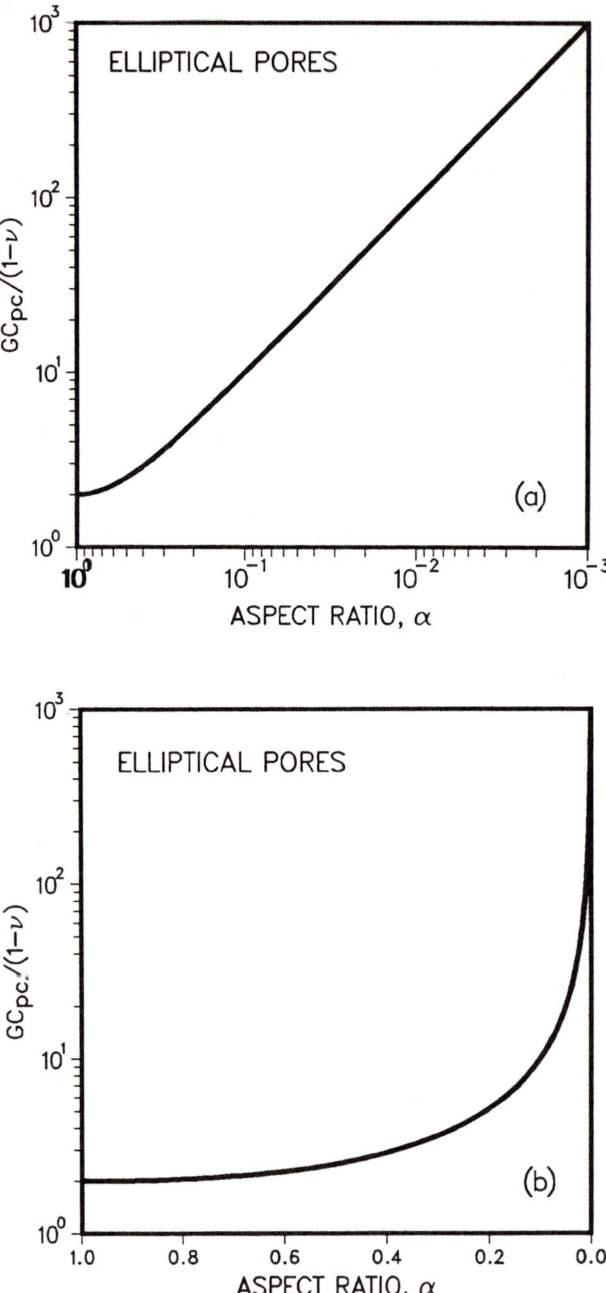

Fig. 8.2. Plane-strain pore compressibility C_{pc} of a tubular pore of elliptical cross-section, as a function of aspect ratio, in both log-log (a) and semi-log (b) format, according to eqn. (8.4).

Slight eccentricity therefore does not markedly effect the pore compressibility, since the influence of e is second-order. This can be seen in Fig. 8.2b, where the slope of the C_{pc} *vs.* α curve is zero near $\alpha=1$. For example, an elliptical hole with an aspect ratio of 0.9 would have an eccentricity of 10%, but its pore compressibility C_{pc} would be only 0.5% greater than that of a circular hole. The circular hole is the "stiffest" type of elliptical hole, and the pore compressibility of elliptical pores increases monotonically as the aspect ratio decreases. As the aspect ratio gets very small, the term $1/\alpha$ dominates eqn. (8.4), and C_{pc} increases as $1/\alpha$ (see Fig. 8.2a).

Although ellipses have been widely used as models for pores in sandstones, they are obviously an oversimplification of the shape of real pores. The compressibility of cylindrical pores of other cross-sectional shapes can be found by using the complex-variable methods developed by Kolosov [1909] and Muskhelishvili [1953]. These methods are based on the fact that solutions to the two-dimensional elasticity equations can be expressed in terms of a pair of complex-valued potentials. By utilizing various results from analytic function theory, the potentials for a given problem can be found from the applied loads. Many specific solutions for the compression of bodies containing two-dimensional cavities of various shapes are discussed by Savin [1961, pp. 47-54]. While Savin was concerned mainly with stress concentrations around the boundaries of the holes, the compressibilities can be found from his solutions by integrating the displacements around the perimeters of the holes. A general method of determining C_{pc} for two-dimensional cavities of known but essentially arbitrary shape has been developed by Zimmerman [1986]. The details of the Kolosov-Muskhelishvili methods used in this analysis are actually quite involved, and their derivation can be found in the text of Sokolnikoff [1956]. The results of the analysis, however, can be presented without any reference to the details of the solution.

Consider a hole Γ of a prescribed but not necessarily regular shape, plotted on a complex z-plane (Fig. 8.3). As long as the hole shape is not mathematically pathological, it is known [Kantorovich and Krylov, 1958] that there will exist a unique mapping that transforms the unit circle γ in the ζ-plane into the hole Γ in the z-plane. This mapping will have the form

$$z(\zeta) = \frac{1}{\zeta} + \sum_{n=1}^{\infty} a_n \zeta^n \,, \tag{8.7}$$

where the a_n are complex constants. This mapping transforms the region outside of the hole Γ into the region inside of the unit circle γ, and, in particular, maps the point at infinity in the physical z-plane into the origin in the transformed ζ-plane. Although it is no simple matter to find the a_n coefficients for a given irregular shape, the Weierstrass theorem assures that these coefficients exist and are unique.

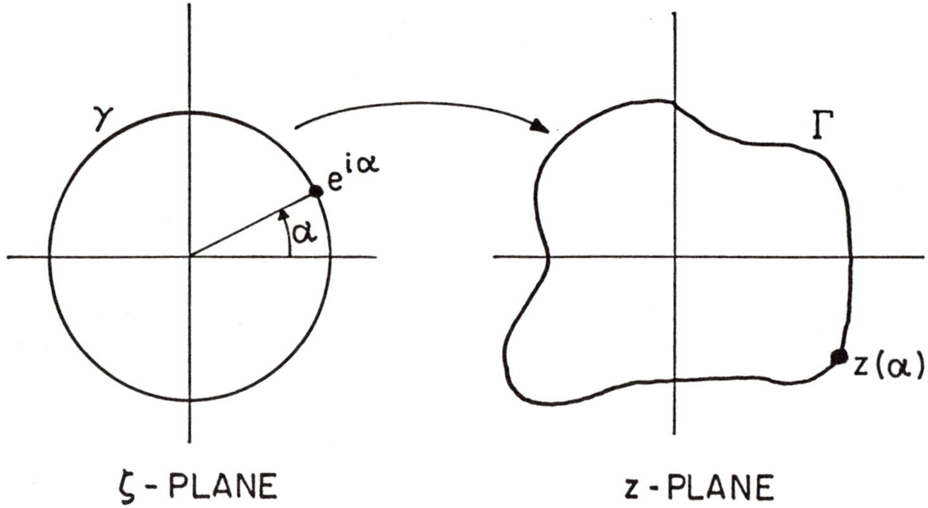

Fig. 8.3. Mapping of the interior of the unit circle in the ζ–plane into the physical region exterior to the hole Γ in the z–plane, according to eqn. (8.7).

By solving the elastostatic boundary value problem in which there is uniform pressure over the boundary Γ, and zero stresses infinitely far from the hole, Zimmerman [1986] found the following closed-form expression for the plane-strain pore compressibility:

$$C_{pc} = \frac{2(1-\nu)}{G} \; \frac{1 + \sum_{n=1}^{\infty} n\,|a_n|^2}{1 - \sum_{n=1}^{\infty} n\,|a_n|^2}\,, \tag{8.8}$$

where $|a_n| = [\mathrm{Re}(a_n)^2 + \mathrm{Im}(a_n)^2]^{1/2}$ is the magnitude of the complex number a_n. Actually, there are some mathematical difficulties that must be considered for cases where the series in the mapping function extend to $n = \infty$. In practice this is not of major concern, since a given shape can always be mapped with arbitrarily small error by using some finite number of terms in the mapping function. Furthermore, it can be shown that while the series that arise in the expressions for the stresses may diverge for some mapping functions, which merely reflects the infinite stress concentrations that are present at sharp corners, the series that appears in eqn. (8.8) for C_{pc} always converges. This is plausible from a physical point of view, since while the stresses may locally be unbounded at certain singular points such as sharp corners, the total areal closure of the hole should always be finite.

There are many interesting facts that can be gained from eqn. (8.8), without considering specific pore shapes. First, it seems that C_{pc} is always proportional to $2(1-\nu)/G$, as was shown in eqn. (8.4) for the special case of elliptical pores. Interestingly, then, the pore compressibility is always more closely related to the shear modulus G than to the bulk modulus K. As an extreme example, a pore in an incompressible ($K=\infty$) but non-rigid ($G<\infty$) material will have a non-zero compressibility. All of the a_n coefficients in the mapping function (8.7) will be zero for a circular hole, so these coefficients reflect the departure of the hole shape from circularity. Since the terms $|a_n|^2$ are all necessarily positive, the pore compressibility of any non-circular pore must be greater than that of a circular pore.

One simple family of mapping functions that can represent a wide variety of shapes is the hypotrochoid, which has the form

$$z(\zeta) = \frac{1}{\zeta} + m\zeta^n \,, \tag{8.9}$$

in which n is a positive integer, and m is a real number that must satisfy $0 \le m < 1/n$ in order for the mapping to be single-valued. A hypotrochoid of exponent n has $n+1$ "corners", which become more acute as m increases. As $m \to 0$ the hole rounds off into a circle, while the corners become cusp-like as $m \to 1/n$. A family of such curves is shown in Fig. 8.4 for the case $n = 3$, with each curve labelled by its value of n, as well as its pore compressibility. Note that if the circle γ in the ζ-plane is parametrized by $\zeta = e^{i\alpha}$, then the hole Γ in the physical z-plane will be described by $e^{-i\alpha} + me^{in\alpha}$. Eqn. (8.8) shows that the pore compressibility of a hypotrochoid is given by [Mavko, 1980]

$$\frac{C_{pc}}{C_{pc}^{o}} = \frac{1+nm^2}{1-nm^2} \,, \tag{8.10}$$

where $C_{pc}^{o} = 2(1-\nu)/G$ is the pore compressibility of a circular hole.

A four-sided hypotrochoid such as that shown in Fig. 8.4 could be used as a simplified two-dimensional model for the compressibility changes that occur as a result of the diagenetic alteration of a sandstone. When $m = 1/3$, the pore resembles to a certain extent a cross-section of the space between a set of four adjacent spherical grains. As m decreases to zero, the shape of the pore will mimic the pore shape evolution undergone by a sandstone as it undergoes diagenesis. Although this model has not yet been used to study the compaction of sandstones, it has been used to study the similar problem of the sintering of powdered metals [O'Donnell and Steif, 1989].

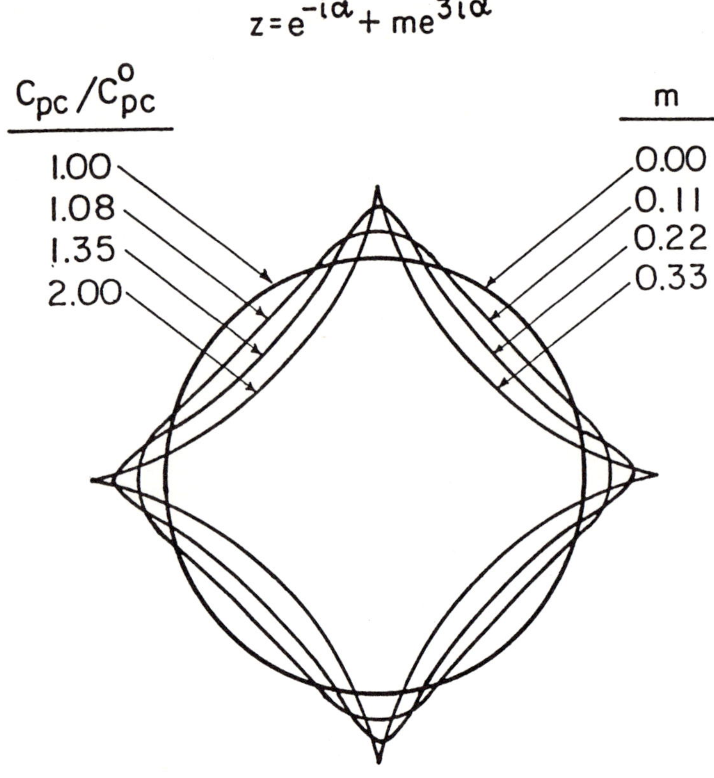

Fig. 8.4. Compressibility of a four-sided hypotrochoidal pore, normalized with respect to that of a circular pore. The equation maps the unit circle in the ζ–plane into the pore boundary in the physical z–plane.

When $m = 2/n\,(n+1)$, the hypotrochoid represents the first two terms of the Schwarz-Christoffel mapping for an equilateral polygon of $n+1$ sides [Savin, 1961], and resembles a polygon with slightly concave sides and rounded corners. The two series in eqn. (8.8) then contain only the term $[2/n\,(n+1)]^2 n$, so that as a function of the number of sides s,

$$\frac{C_{pc}}{C_{pc}^{o}} = \frac{s^2(s-1)+4}{s^2(s-1)-4} \, . \tag{8.11}$$

It follows from eqn. (8.11) that the quasi-triangular pore is 57% more compressible than a circular pore, and the quasi-square only 18% more compressible. The compressibility of these pores decreases rapidly as the number of sides increases and the pore becomes more circular.

The series in eqn. (8.8) typically converge very rapidly, so that the compressibility of a particular pore can usually be well approximated using only a few terms in its mapping series. For example, consider the complete mapping function for the square hole [Savin, 1961]:

$$z(\zeta) = \frac{1}{\zeta} + \frac{1}{6}\zeta^3 + \frac{1}{56}\zeta^7 + \frac{1}{176}\zeta^{11} + \frac{1}{384}\zeta^{15} + \ldots \tag{8.12}$$

The compressibility of the two-term quasi-square, normalized with respect to that of a circle, is 1.182. The addition of the remainder of the terms in the mapping function, which causes the hole to more closely approximate a square, increases the normalized pore compressibility to only 1.188. The rate at which the C_{pc} values of the approximate shapes converge to that of the square is illustrated in Fig. 8.5. This rapid convergence is useful for holes of complicated shapes, for which the mapping functions must be found through laborious numerical algorithms [Kantorovich and Krylov, 1958].

Although eqn. (8.8) expresses the compressibility explicitly in terms of the mapping function, it is worthwhile to try to relate the compressibility to more geometrically meaningful parameters such as the perimeter of the hole, Π. First note that C_{pc} can be thought of as the constant of proportionality relating the area change (under unit confining pressure) to the initial area. One way to view eqn. (8.8), then, is to observe that since C_{pc} depends on hole shape, the area change does not correlate closely with the initial area, A^i. Surprisingly, however, for many different hole shapes, the change in the area of the pore that is caused by a given increment of confining pressure correlates extremely closely with the square of the perimeter. To see this, divide the normalized compressibility C_{pc}/C_{pc}^o by $\Pi^2/4\pi A^i$, a dimensionless ''specific surface'' parameter which equals unity for a circle, and exceeds unity for all other shapes. This new compressibility measure β_{pc} in effect normalizes the area change due to a confining pressure increment against Π^2, since

$$\beta_{pc} = \frac{C_{pc}}{C_{pc}^o}\frac{4\pi A^i}{\Pi^2} = \frac{2\pi G\,\Delta A}{(1-\nu)\,\Pi^2\,\Delta P},$$

so

$$\frac{\Delta A}{\Delta P} = \frac{(1-\nu)}{2\pi G}\,\beta_{pc}\,\Pi^2\,. \tag{8.13}$$

This normalized pore compressibility measure β_{pc} has some similarity to the ''hydraulic

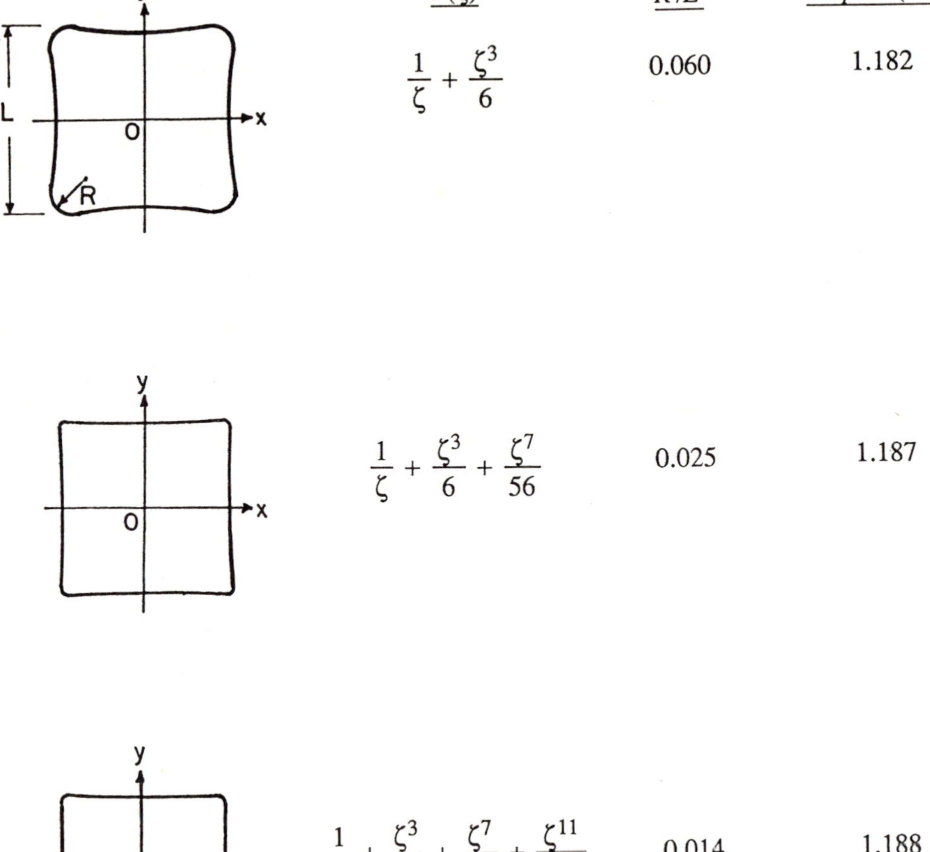

Fig. 8.5. Successive approximations to a square hole, showing the shape, the mapping function, the radius of curvature at the corner, and the value of C_{pc}. An infinite number of terms would be needed for a truly square hole with sharp corners; its normalized pore compressibility would be 1.188, the same as that of the four-term approximation.

diameter'' concept used in fluid mechanics [White, 1974], which states that the hydraulic resistance of a pipe of a given cross-section is roughly proportional to the square of its wetted perimeter.

Evaluation of eqn. (8.13) for a variety of hole shapes shows that β_{pc} does not deviate appreciably from unity, since both C_{pc} and $\Pi^2/4\pi A^i$ increase as the hole shape departs from circularity. In general, determination of Π for a given mapping function will require numerical evaluation of the following integral:

$$\Pi = \int_{\gamma} |dz| = \int_0^{2\pi} |z'(\alpha)| d\alpha . \tag{8.14}$$

For the hypotrochoid, however, the above integral can be evaluated in terms of elliptic integrals. First, put $\zeta = e^{i\alpha}$ on γ, in which case $z(\alpha) = e^{-i\alpha} + me^{in\alpha}$, and $z'(\alpha) = -ie^{-i\alpha} + imne^{in\alpha}$. Insertion of this expression for $z'(\alpha)$ into eqn. (8.14) leads to

$$\Pi = \int_0^{2\pi} [1+m^2n^2+2mn\cos(n-1)\alpha]^{1/2} d\alpha$$

$$= (1+m^2n^2)^{1/2} \int_0^{2\pi} \left[1+\frac{2mn}{1+m^2n^2}\cos(n-1)\alpha\right]^{1/2} d\alpha$$

$$= (1+m^2n^2)^{1/2} \int_0^{2\pi} \left[1+\frac{2mn}{1+m^2n^2}\cos\theta\right]^{1/2} d\theta$$

$$= (1+m^2n^2)^{1/2} \int_0^{2\pi} \left[1+\frac{2mn}{1+m^2n^2}\left(1-2\sin^2\frac{\theta}{2}\right)\right]^{1/2} d\theta$$

$$= (1+m^2n^2+2mn)^{1/2} \int_0^{2\pi} \left[1-\frac{4mn}{1+m^2n^2+2mn}\sin^2\frac{\theta}{2}\right]^{1/2} d\theta$$

$$= (1+mn) \int_0^{2\pi} \left[1-\frac{4mn}{(1+mn)^2}\sin^2\theta\right]^{1/2} d\theta$$

$$= 4(1+mn)\int_0^{\pi/2}\left[1-\frac{4mn}{(1+mn)^2}\sin^2\theta\right]^{1/2}d\theta$$

$$= 4(1+mn)E\left[\frac{2\sqrt{mn}}{1+mn}\right], \tag{8.15}$$

where E is the complete elliptic integral of the second kind.

Since the initial area of the hypotrochoid can be shown [Zimmerman, 1986] to equal $\pi(1-nm^2)$, eqns. (8.10), (8.13) and (8.15) show that

$$\beta_{pc} = \frac{4\pi A^i}{\Pi^2}\frac{1+nm^2}{1-nm^2}$$

$$= \frac{4\pi^2(1-nm^2)(1+nm^2)}{16(1+mn)^2(1-nm^2)E^2[2\sqrt{mn}/(1+mn)]}$$

$$= \frac{\pi^2(1+nm^2)}{4(1+mn)^2E^2[2\sqrt{mn}/(1+mn)]}, \tag{8.16}$$

where $E^2(x)$ denotes $[E(x)]^2$. When $m=0$, the hypotrochoid degenerates into a circle; since $E(0)=\pi/2$, eqn. (8.16) shows that $\beta_{pc}=1$, as expected. For any fixed value of n, the maximum allowable value of m is $1/n$, for which the argument of the elliptical integral becomes unity. Since $E(1)=1$, the value of β_{pc} for $m=1/n$ will be $\pi^2(n+1)/16n = 0.617(n+1)/n$. In terms of the number of sides of the hypotrochoid, this can be written as $0.617s/(s-1)$. The factor β_{pc} therefore lies between 0.926 and 1.00 for a three-sided hypotrochoid, for example, and between 0.823 and 1.00 for a four-sided hypotrochoid. For the quasi-polygons obtained by setting $m=2/n(n+1)$ in eqn. (8.9), β_{pc} always lies in the range 0.95 – 1.00. This phenomenon is by no means restricted to holes which are roughly equi-dimensional; for the infinitely thin elliptical slit obtained by setting n=1 and letting $m\rightarrow 1$ in eqn. (8.9), which is discussed in more detail in Chapter 9 as a model for partially-contacting grain boundaries, the value of β_{pc} is only 1.23. Fig. 8.6 illustrates this situation for a few different shapes, such as a circle, square, triangle, rectangular slit, and thin ellipse. Note that the parameter β_{pc} is on the order of unity for all of these shapes. If this relation $\beta_{pc}=1$ were precisely true, C_{pc} would then always equal $(1-\nu)\Pi^2/2\pi GA^i$. The fact that β_{pc} always seems to be close to unity raises the possibility of using computerized image analysis to estimate Π and A^i from thin samples, thereby arriving at an estimate of C_{pc} for an actual non-idealized sandstone pore geometry.

SHAPE	$GC_{pc}/2(1-\nu)$	$P^2/4\pi A$	$GC_{pc}/2(1-\nu)\cdot 4\pi A/P^2$
○	1.000	1.000	1.000
△	1.581	1.654	0.956
□	1.188	1.273	0.933
▭	$0.393\,R^{-1}$	$0.318\,R^{-1}$	1.234
⬭	$0.500\,R^{-1}$	$0.405\,R^{-1}$	1.234

R = ASPECT RATIO

Fig. 8.6. Pore compressibilities of two-dimensional pores of various cross-sections. The first column shows the shape, the second column lists C_{pc} normalized with respect to that of a circular hole, the third column lists the normalized "specific surface" parameter, and the fourth column lists the parameter β_{pc} [see eqn. (8.13)]. Note that in slight contrast to the notation used in the remainder of the book, in this figure the perimeter is denoted by P instead of Γ, and the aspect ratio is denoted by R instead of α.

Chapter 9. Two-Dimensional Cracks

The models discussed in Chapter 8 are intended to apply to the roughly equi-dimensional pore bodies that exist in the regions between groups of nearby grains. However, not all of the void spaces in a sandstone fit into this category; there are also very thin ''crack-like'' pores that exist along grain-boundaries. As the diagenetic process proceeds, these intergranular voids tend to disappear as they become filled with cementing material and become welded together through pressure solution, etc. Pore casts of the void spaces of sandstones show [Pittman, 1984] that few grain boundaries are completely bonded. These grain-boundary pores are characterized by having a characteristic thickness in one dimension that is much smaller than in the other two. In principle, the compressibility of such pores can be found using the conformal-mapping method discussed in Chapter 8. Many different types of crack-like pores have been analyzed by various other approaches within the theory of elasticity, however, and so it is convenient to discuss them on an *ad hoc* manner, following their initial treatment.

The elliptical crack, which was first studied by Inglis [1913] as a model for flaws in structural steel, is one of the most widely used models for crack-like voids in rocks. Griffith [1920] used the Inglis solution in his classic paper on rupture, and since then the thin elliptical crack has often been referred to as the Griffith crack. The normal displacement at the crack surface which results from this solution was integrated by Walsh *et al.* [1965], and the resulting pore compressibility is given by eqn. (8.4). This result can also be derived from eqn. (8.10) by taking $n = 1$ and $m = (1-\alpha)/(1+\alpha)$. For the very low aspect ratios which characterize grain-boundary voids, this reduces to (see Fig. 8.2)

$$C_{pc} = \frac{(1-\nu)}{\alpha G} \quad \text{for small } \alpha \,. \tag{9.1}$$

Note that an ''infinitely thin'' crack has an infinite compressibility; this is not to say that the volume change for a thin crack will be infinitely large, but merely reflects the fact that the definition of C_{pc} includes the term $1/V_p^i$, and the *initial* volume of an infinitely thin crack is zero.

The property that cracks have which leads to nonlinearities in the stress-strain curve is that they close up under suitably large effective confining stresses, after which they cease to contribute to the compressibility. The actual rate at which they close, as a function of the effective confining stress, is an important component in the resulting stress-strain relations. Eqn. (9.1) gives the instantaneous compressibility of a crack of aspect ratio α. As a crack closes, however, its aspect ratio will decrease, and it is not *a priori* clear how to interpret eqn. (9.1) in this case. Sneddon and Elliott [1946] examined the deformation of such a crack in detail, and showed that, to a very high approximation, the length of the crack remains unchanged during closure, the aperture of the crack decreases linearly with pressure, and the shape remains essentially elliptical. As can be seen from eqn. (5.7), the difference between C_{pc} and C_{pp} is negligible for cracks, since this difference is an order of magnitude in α smaller than either C_{pp} or C_{pc}. Consequently, the compression of a crack depends only on the differential pressure P_d, which, for the remainder of this section, will be denoted simply by P. For a crack of unit half-length, and initial half-width $w^i = \alpha$, Sneddon and Elliott showed that the instantaneous half-width is related to the pressure by $w = w^i - (1-\nu)P/G$ (Fig. 9.1). The open area of the crack will always equal πw, since the deformed shape is still elliptical, and so the open area is related to the pressure by (see Fig. 9.1)

$$\frac{A(P)}{A^i} = 1 - \frac{(1-\nu)P}{\alpha G} , \tag{9.2}$$

in which α is always to be interpreted as the *initial* aspect ratio of the crack. The open area of the crack is a linear function of the pressure P, until the pressure is reached at which it closes up entirely. As long as the crack is open, its compressibility is therefore given by $(1-\nu)/\alpha G$. The pressure P^* at which the crack closes is found by setting $A(P)=0$, which yields

$$P^* = \frac{\alpha G}{(1-\nu)} . \tag{9.3}$$

Hence the pore compressibility C_{pc} is equal to $(1-\nu)/\alpha G$ when $P < P^*$, but equals 0 when $P > P^*$. This type of linear closure is the simplest possible, and it would be convenient if this closure rate could serve as a reasonable approximation for all cracks. The extent to which this is practical can be assessed by examining cracks of different shapes.

One type of idealized crack shape that is very different from the Griffith crack is a thin rectangular slit which has absolutely no rounding off at its edges (Fig. 9.2). The compression of such a crack has been studied by Berry [1960], who used the rectangular slit as a model for an underground mine seam. Consider such a slit-like crack with an initial length arbitrarily put equal to 2, and initial width 2α. The vertical displacement at point x on the face of the crack can be denoted by $w(x)$. For pressures low enough so that the opposing faces of the crack are

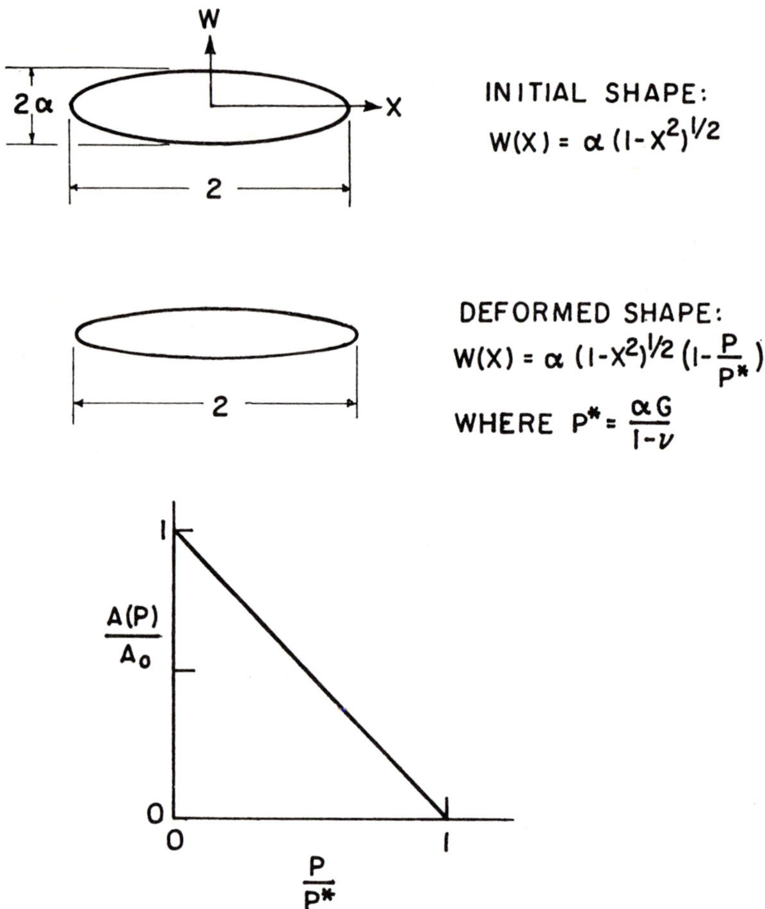

Fig. 9.1. Closure of a two-dimensional elliptical crack. The open area is a linear function of pressure, according to eqn. (9.2), until the crack closes completely at a pressure of $P^* = \alpha G/(1-\nu)$.

not in contact, Berry found that

$$w(x) = \frac{(1-\nu)(1-x^2)^{1/2}P}{G} \, . \tag{9.4}$$

Initially, then, the faces of this crack deform into "concave" ellipses. At some critical touching pressure P_t, the two faces of the crack meet at their midpoints. The value of P_t is found by setting $w(0)=\alpha$, from which it follows that $P_t = \alpha G/(1-\nu)$, which is precisely equal to the *closing* pressure for the Griffith crack. For pressures below this value, $\Delta A = -\pi w(0)$, which is the area of the elliptical indentations, so that $\Delta A = -\pi(1-\nu)P/G$. Hence for pressures below the touching pressure, the open area is a linear function of the pressure, and $C_{pc} = \pi(1-\nu)/4\alpha G$, which is 21% less than for a Griffith crack. This difference is due entirely to the differences in the initial areas of the rectangular and elliptical cracks, which appear in the denominator of C_{pc}. The rate at which the open area of the cracks decreases with pressure is actually the same for the two cracks: $dA/dP = -\pi(1-\nu)/G$.

The character of the solution changes after the sides make contact, as depicted in Fig. 9.2. For $P \geq \alpha G/(1-\nu)$, the faces are in contact over some region $|x| \leq b$. Assuming that the crack faces are frictionless, the length of the closed region can be related to the pressure through the relation

$$\frac{P^*}{P} = E\left[\sqrt{(1-b^2)}\right] - b^2 K\left[\sqrt{(1-b^2)}\right] , \tag{9.5}$$

where $\{K, E\}$ are the complete elliptic integrals of the first and second kind, respectively. The shapes of the crack faces in the regions $b \leq |x| \leq 1$ are still elliptical, so the following expression for the open area of the crack as a function of the closure parameter b holds:

$$\frac{A(P)}{A^i} = \left[1 - \frac{\pi}{4}\right](1-b). \tag{9.6}$$

With the aid of tables of elliptic integrals [Fletcher, 1940], eqns. (9.5) and (9.6) can be solved numerically for $A(P)/A^i$. The resulting curve for $A(P)/A^i$ shown in Fig. 9.2 is not very different from the linear closure found for the elliptical crack. For example, when $P = P^* = \alpha G/(1-\nu)$, the elliptical crack would be completely closed, while the "rectangular" crack will be 79% closed. This is the pressure at which the crack faces first touch, after which the closure rate drops off severely. When P reaches $2P^*$, closure will be 93% complete. The open cross-sectional area of this crack never completely vanishes, since the vertical displacement at the edges (where $x = \pm 1$) is always zero.

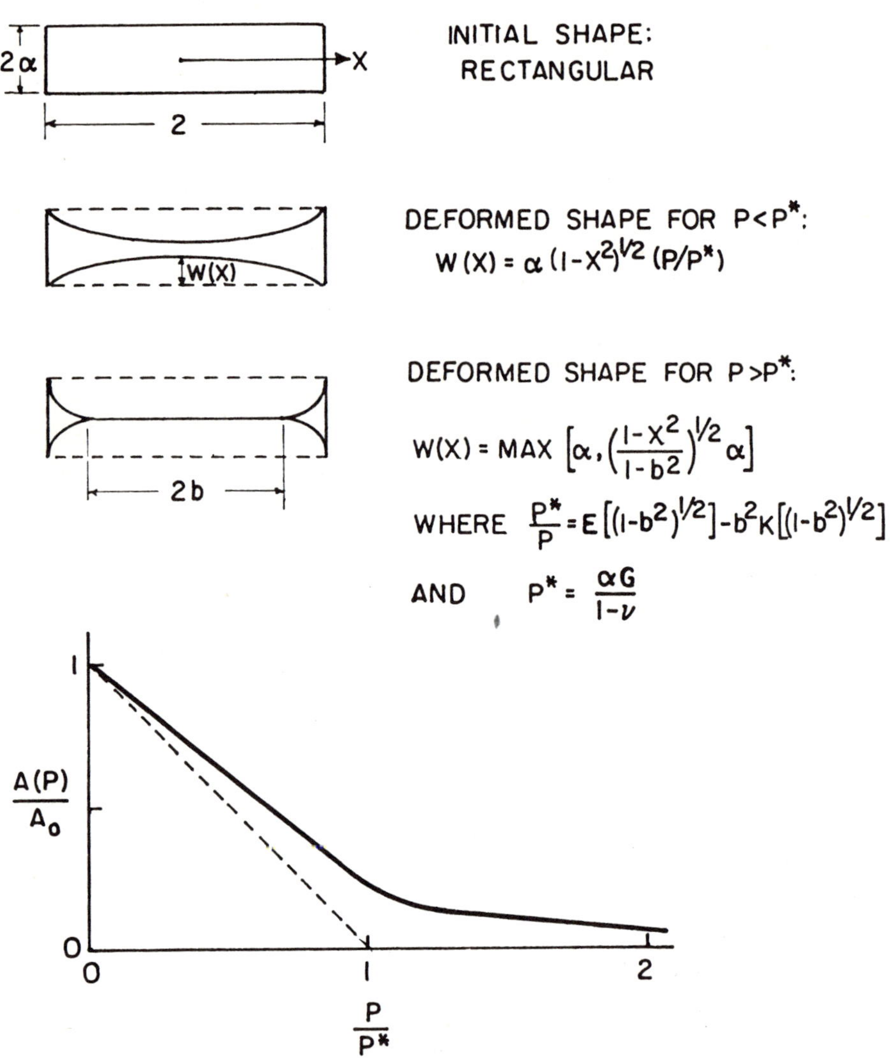

Fig. 9.2. Closure of a two-dimensional slit-like crack (after Berry [1960]). The deformation is described by eqns. (9.4-9.6).

The rectangular slit model for cracks is unrealistic in at least one regard, however, since a crack of this shape would have infinite stress concentrations at the corners. In fact, this is also a problem with the infinitely thin elliptical crack. Barenblatt [1962] pointed out that cracks with such infinite stress concentrations would not actually exist, and showed that only cracks with smoothly tapered edges are thermodynamically stable. A general method for dealing with cracks which have tapered edges, but otherwise arbitrary profiles, was developed by Mavko and Nur [1978], using elastic dislocation theory. They applied their method to the compression of two different crack shapes with tapered edges. In light of the scarcity of such exact solutions for the compression of cracks, these solutions are worth examining in detail.

Consider first the crack shown in Fig. 9.3, which has an initial length of 2, and an initial width at its midpoint of 2α. The initial profile of this crack is given by

$$w^i(x) = \alpha(1-x^2)^{3/2} . \tag{9.7}$$

Under a differential pressure P, the crack assumes a shape in which the edges are closed for $|x| \geq c$, while the profile for the open part of the crack is given by

$$w(x) = \alpha(c^2-x^2)^{3/2} , \tag{9.8}$$

where

$$c^2 = 1 - \frac{2(1-\nu)P}{3\alpha G} . \tag{9.9}$$

The pressure needed to close the crack is found by setting $w(0)=0$, from which it follows that $c=0$, and $P[\text{closure}] = 3\alpha G/2(1-\nu)$. This is 50% higher than the closing pressure for the elliptical crack, reflecting the stiffening effect as the cusp-shaped edges come together. Mavko and Nur [1978] actually proved the extremely important result that the non-normalized crack-closure rate dA/dP of *any* two-dimensional crack with tapered edges depends only on the length of the crack, and equals $-\pi(1-\nu)/G$. The stiffening of a crack as it is compressed is due to the decrease in the crack length as the two faces come together. The complete pressure-area relation for the crack shown in Fig. 9.3 is found by integrating $w(x)$ across the length of the crack, and using eqn. (9.9) to relate the instantaneous half-length c of the crack to the pressure P:

$$A(P) = 2\int_{-1}^{+1} w(x)dx = 2\int_{-c}^{+c} \alpha c^3\left[1-(x/c)^2\right]^{3/2} dx = \frac{3\pi\alpha c^4}{4} . \tag{9.10}$$

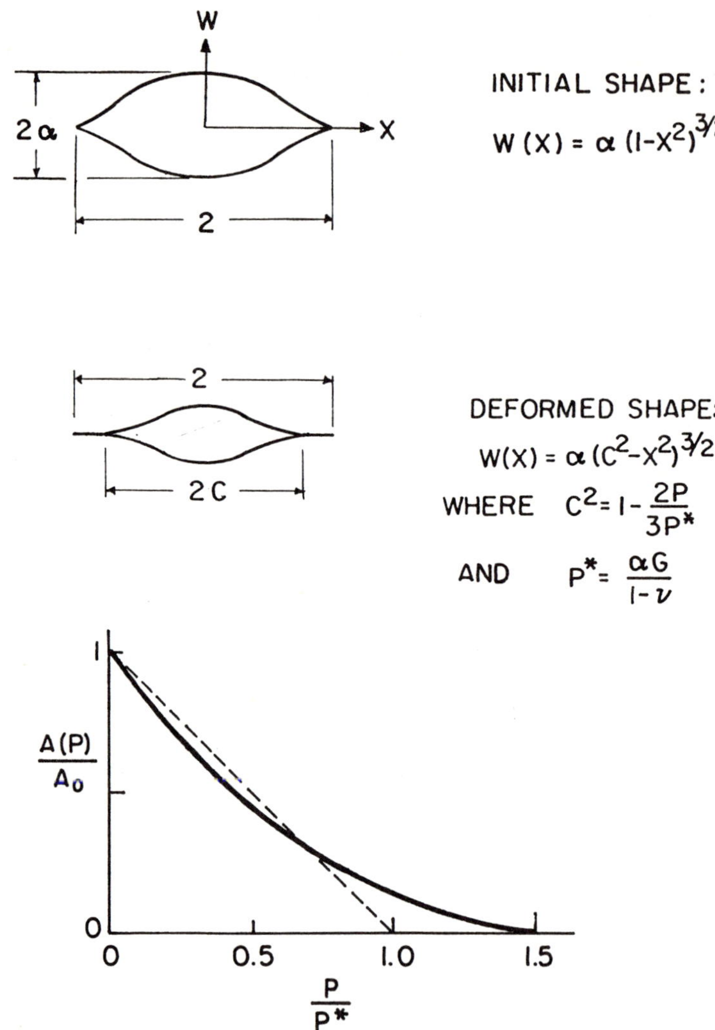

Fig. 9.3. Closure of the two-dimensional cusp-like crack whose initial profile is given by eqn. (9.7) (after Mavko and Nur [1978]). The deformation is described by eqns. (9.8-9.11).

Since $c = 1$ when $P = 0$, $A(P)/A^i = c^4$, which when combined with eqn. (9.9) yields (after eliminating c)

$$\frac{A(P)}{A^i} = \left[1 - \frac{2(1-\nu)P}{3\alpha G}\right]^2 . \tag{9.11}$$

Although the closure curve of this crack is not a linear function of the pressure (Fig. 9.3), it straddles the closure curve for the elliptical crack fairly closely. At the elliptical crack closing pressure of $\alpha G/(1-\nu)$, this tapered crack will be 89% closed. In fact, the difference in the open areas of the two cracks will never be greater than 11%, at any pressure.

The other crack (Fig. 9.4) treated in detail by Mavko and Nur [1978] has an initial profile given by

$$w^i(x) = \alpha(1 - 0.3x^2 - 1.3x^4)(1 - x^2)^{1/2} . \tag{9.12}$$

This crack also shortens as it closes up, and has the property that its two faces touch each other at a pressure equal to $1.12\alpha G/(1-\nu)$. Determining the shape of the crack after this initial contact would be extremely difficult, but for pressures below this value the crack profile is given by

$$w(x) = \frac{-\alpha}{41}\left[54x^4 + (27c^2 - 40)x^2 + (40c^2 - 81c^4)\right](c^2 - x^2)^{1/2} , \tag{9.13}$$

where P is related to the instantaneous half-length c by

$$P = \frac{\alpha G}{(1-\nu)}\frac{10}{328}\left[33 + 48c^2 - 81c^4\right] . \tag{9.14}$$

Since P is the independent variable, it is convenient to invert this relation, which is merely a quadratic in c^2, to find

$$c = \left\{0.296 + \left[0.495 - 0.405(P/P^*)\right]^{1/2}\right\}^{1/2} , \tag{9.15}$$

where $P^* = \alpha G/(1-\nu)$.

The area-pressure relationship is found as before, by integrating the aperture function across the length of the crack. The integral is evaluated by making the change of variable

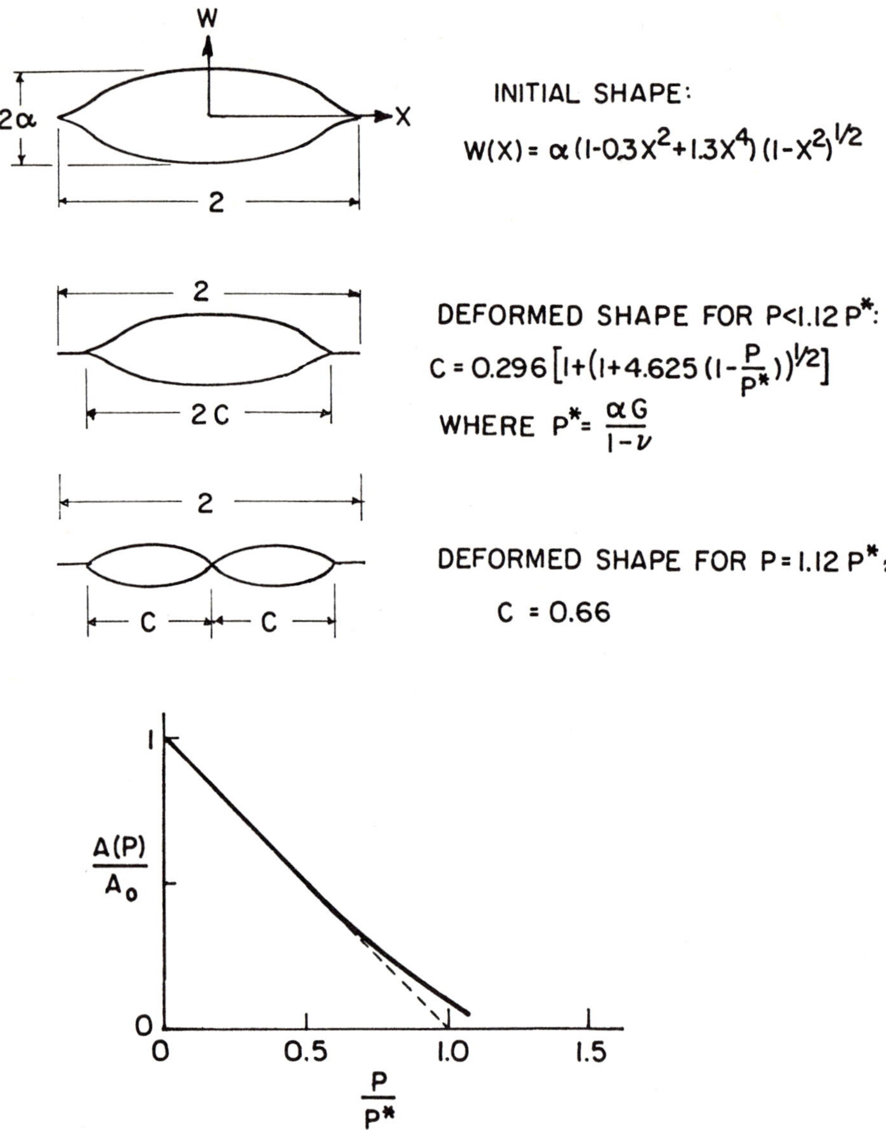

Fig. 9.4. Closure of the two-dimensional cusp-like crack whose initial profile is given by eqn. (9.12) (after Mavko and Nur [1978]). The deformation is described by eqns. (9.13-9.17).

$x = c\ \sin t$,

which reduces the integrand to a sum of trigonometric functions:

$$A = \frac{-2\alpha}{41}\int_{-c}^{+c}[54x^4 + (27c^2-40)x^2 + (40c^2-81c^4)](c^2-x^2)^{1/2}dx$$

$$= \frac{-2\alpha c^2}{41}\int_{-\pi/2}^{+\pi/2}[54c^4\sin^4 t + (27c^2-40)c^2\sin^2 t + (40c^2-81c^4)]\cos^2 t\,dt$$

$$= \frac{-2\alpha c^2}{41}\left[54c^4\frac{\pi}{16} + (27c^2-40)c^2\frac{\pi}{8} + (40c^2-81c^4)\frac{\pi}{2}\right]$$

$$= \frac{15\pi\alpha c^4}{82}(9c^2-4)\ . \qquad (9.16)$$

Since the initial area corresponds to $c=1$, $A^i = 75\pi\alpha/82$, and so

$$\frac{A(P)}{A^i} = \frac{c^4(9c^2-4)}{5}\ . \qquad (9.17)$$

Taken together, eqns. (9.15) and (9.17) give the open area as an explicit function of the pressure. The closure curve is plotted in Fig. 9.4, where it is seen that this crack follows the elliptical crack closure curve extremely closely. For example, when $P = 0.5P^*$, this tapered crack will be 50.4% closed, while at $P = P^*$ the closure will be 95.4%. At a certain pressure, the faces of the crack will touch at their midpoint. This pressure can be found by first setting $w(x=0)=0$ in eqn. (9.13), which yields $c^2=40/81$. The pressure necessary to cause this amount of crack-shortening is found from eqn. (9.14) to be $1.12P^*$. At this pressure, eqn. (9.17) shows that the crack is already 97.7% closed.

The rate at which "cracks" close under pressure determines the shape of the stress-strain curve of a rock. The analysis given in this chapter shows that the pressure needed to close up a two-dimensional crack is roughly equal to $\alpha G/(1-\nu)$, regardless of the precise shape of the crack, as long as the aspect ratio α is defined in the obvious way. In Chapter 10, it will be seen that this result also holds for three-dimensional cracks, aside from a multiplicative constant on the order of unity. In Chapter 12, distributions of aspect ratios will be used to model the complete elastic portion of the hydrostatic stress-strain curve of a sandstone.

Crack closure rates not only have a major influence on mechanical properties such as compressibility and wave velocities, but also influence transport properties such as permeability, electrical conductivity, and thermal conductivity. Each of these transport properties decreases with increasing confining pressure, although the degree of the stress-dependence is usually not as strong as it is for compressibility. (An exception to this are tight gas sands, whose permeabilities may decrease by as much as 90% [Ostensen, 1983] as the differential stress increases from zero to a few thousand psi.) Brower and Morrow [1985] used crack shapes similar to those treated in this chapter to model the stress-dependence of permeability in tight gas sands. Yale [1984] modeled the pore space in sandstones as a network of tubular pores whose cross-sections are elliptical or cusp-shaped (as in the analysis of Mavko and Nur [1978]), and calculated the variation of permeability and electrical resistivity with stress. The pressure-dependence of the thermal conductivity of fluid-saturated rocks was studied by Walsh and Decker [1966], using the penny-shaped crack as a pore model (see Chapter 10).

Chapter 10. Spheroidal Pores

The idealized pore shapes discussed in Chapters 8 and 9 were all two-dimensional models of sandstone pores which in reality are three-dimensional. In two dimensions, the compressibility of a pore of essentially arbitrary shape can be found by using either the conformal mapping method developed by Zimmerman [1986], or the dislocation method developed by Mavko and Nur [1978]. Unfortunately, there are no three-dimensional analogues of either of these two methods, due in the former case to the lack of non-trivial conformal mappings in three-dimensions, and in the latter case to analytical complexities. The only three-dimensional pore shape that is amenable to analytical treatment is the ellipsoid [Sadowsky and Sternberg, 1947,1949]. In its various forms, however, the ellipsoid can represent a variety of pore shapes, such as spheres, cylinders, thin circular cracks, etc. An expression for the pore compressibility of an ellipsoidal pore with three axes of unequal lengths is implicitly contained in the results of Eshelby [1957], who studied the more general problem of elastic inclusions. This result, which was obtained through a sequence of ingenious mathematical manipulations and thought experiments that circumvented the need to solve any boundary-value problems, is unfortunately expressed in terms of relatively unwieldy integrals. Moreover, if the three axes of an ellipsoidal pore each have a different length, two aspect ratio parameters will be needed to describe its geometry. Perhaps, with the increased capabilities of computerized image analysis [Ruzyla, 1986; Berryman, 1989], it may become feasible to model pores as ellipsoids of arbitrary aspect ratios. Currently, however, the spheroid, which is a degenerate ellipsoid having two axes of equal length, is much more widely used as a three-dimensional model for sandstone pores [Warren, 1973; Kuster and Toksöz, 1974; Zimmerman, 1984a].

A spheroidal surface can be thought of as resulting from the revolution of an ellipse about one of its axes of symmetry. Revolution about the minor axis generates an oblate spheroid, while revolution about the major axis produces a prolate spheroid (see Fig. 10.1). For example, a "pancake-shaped" or "doorknob-shaped" pore could be approximated by an oblate spheroid of suitable aspect ratio, while a "cigar-shaped" pore could be approximated by a prolate spheroid. The aspect ratio can be defined as the ratio of the length of the unequal axis to the length of one of the two equal axes. Prolate spheroids therefore have aspect ratios greater than 1.0, while oblate spheroids have aspect ratios less than 1.0. In its two extreme limiting forms, the spheroid degenerates into either an infinitely long needle-like cylinder (as $\alpha \to \infty$), or an

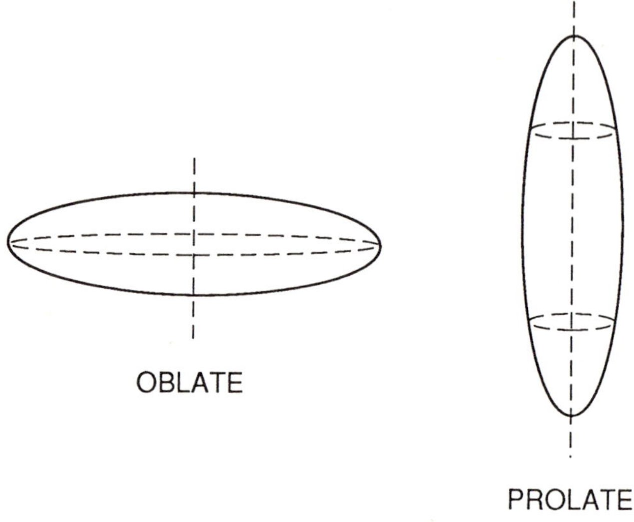

Fig. 10.1. An oblate (left) and a prolate (right) spheroid. In each case, the dotted line represents the axis of revolution.

infinitely thin "penny-shaped" crack (as $\alpha \rightarrow 0$). At the boundary between the oblate and the prolate spheroid is the sphere, whose aspect ratio is 1.0. One advantage of the spheroidal pore model is that a single analytical expression, which is valid over the entire range of aspect ratios, can be found for the pore compressibility. It so happens that it is not convenient to find this expression by examining the limit of the ellipsoidal pore compressibility for the case when two of the axes are of equal length. The pore compressibility of a spheroidal pore is more readily found by solving the appropriate elasticity problem in spheroidal coordinates.

An analytical solution for the volume change of a pressurized spheroidal cavity in an infinite elastic medium was found by Zimmerman [1985a]. As in the case of two-dimensional pores, the pore compressibility was found by solving the elastostatic problem in the region exterior to a spheroidal cavity, with uniform pressure applied at the cavity surface, and zero stresses at infinity. This solution utilized the general solution for the elasticity equations in *prolate* spheroidal coordinates that was derived by Edwards [1951], with the arbitrary coefficients chosen so as to satisfy the boundary conditions. Integrating the normal component of the displacement over the surface of the cavity yields the pore compressibility in the form (see Fig. 10.2):

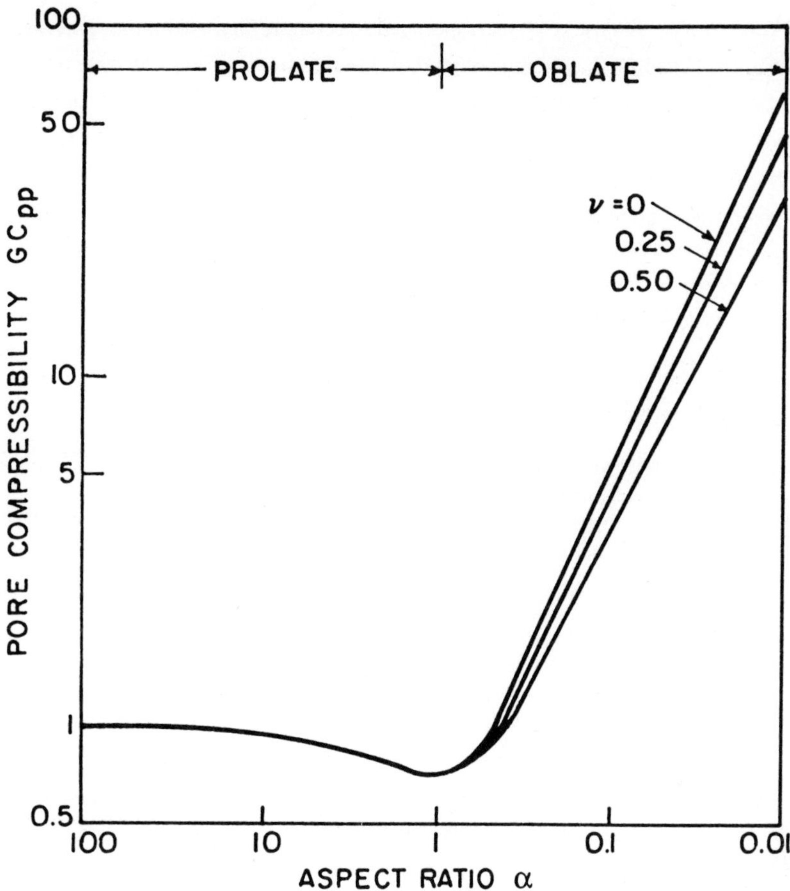

Fig. 10.2. Pore compressibility of a spheroidal pore, as a function of aspect ratio, for various values of the Poisson ratio. Eqn. (10.1) is used for prolate spheroids, and eqn. (10.4) is used for oblate spheroids.

$$GC_{pp} = \frac{2(1-2\nu)(1+2R)-(1+3R)\{[1-2(1-2\nu)R-[3\alpha^2/(\alpha^2-1)]\}}{4\{(1+3R)[\alpha^2/(\alpha^2-1)]-(1+R)(\nu+\nu R+R)\}} ,$$

where $R = \frac{1}{\alpha^2-1} + \frac{\alpha}{2(\alpha^2-1)^{3/2}} \ln\frac{\alpha-\sqrt{(\alpha^2-1)}}{\alpha+\sqrt{(\alpha^2-1)}}$. (10.1)

The pore compressibility C_{pp} is a minimum when the aspect ratio is equal to one, where it has the value $3/4G$. At the other end of the aspect ratio spectrum, the needle-like pore has C_{pp} equal to $1/G$. (In Chapter 8 the infinitely long needle-like pore was treated as a two-dimensional pore of circular cross-section. The present result $C_{pp}=1/G$ can be reconciled with the result $C_{pc}=2(1-\nu)/G$ of Chapter 8 by utilizing the relation $C_{pc}=C_{pp}+C_r$, and recalling that the "two-dimensional" plane-strain compressibility is equal to $C_r/(1-2\nu)$ [Zimmerman, 1986].) The values of C_{pp} for all prolate spheroids lie between $0.75/G$ and $1/G$, and are nearly independent of Poisson's ratio. An asymptotic expression for C_{pp}, whose range of accuracy is roughly $2<\alpha<\infty$, can be found by putting $\nu=0.50$ in eqn. (10.1), and then examining the limit of the expression as $\alpha\rightarrow\infty$, which leads to (see Fig. 10.3)

$$GC_{pp}\approx\frac{2\alpha^2+1}{2\alpha^2+\ln(4\alpha^2)}\quad\text{as}\quad\alpha\rightarrow\infty\,. \tag{10.2}$$

This approximation is valid to within 2% for all $\alpha>2$. A simpler expression, which is accurate to within 2% for all aspect ratios $\alpha>1$, is (Fig. 10.4)

$$GC_{pp}=1-0.25e^{-(\alpha-1)/3}\,. \tag{10.3}$$

Whereas eqn. (10.2) was derived by a rigorous asymptotic expansion [Zimmerman, 1985a], eqn. (10.3) is merely a convenient curve-fit.

The governing differential equations of elasticity in prolate spheroidal coordinates are equivalent to those in oblate spheroidal coordinates, through a certain change of variables [Edwards, 1951]. Rather than solve the governing equations in oblate spheroidal coordinates, it is much easier to merely transform the final expression for C_{pp}. If the pore compressibility for prolate spheroids given by eqn. (10.1) is thusly transformed, the resulting expression for oblate spheroids is [Zimmerman, 1985a]

$$GC_{pp}=\frac{(1+3R)[1-2(1-2\nu)R+3\alpha^2]-2(1-2\nu)(1+2R)}{4\{(1+3R)\alpha^2+(1+R)(\nu+\nu R+R)\}}\,,$$

$$\text{where}\quad R=\frac{-1}{1-\alpha^2}+\frac{\alpha}{(1-\alpha^2)^{3/2}}\arcsin\sqrt{1-\alpha^2}\,. \tag{10.4}$$

Although the expression for the pore compressibility of a a three-dimensional oblate spheroidal pore is much more complicated than that for the two-dimensional elliptical pore, the variation of the compressibility with aspect ratio is very similar (compare Figs. 8.2 and 10.2).

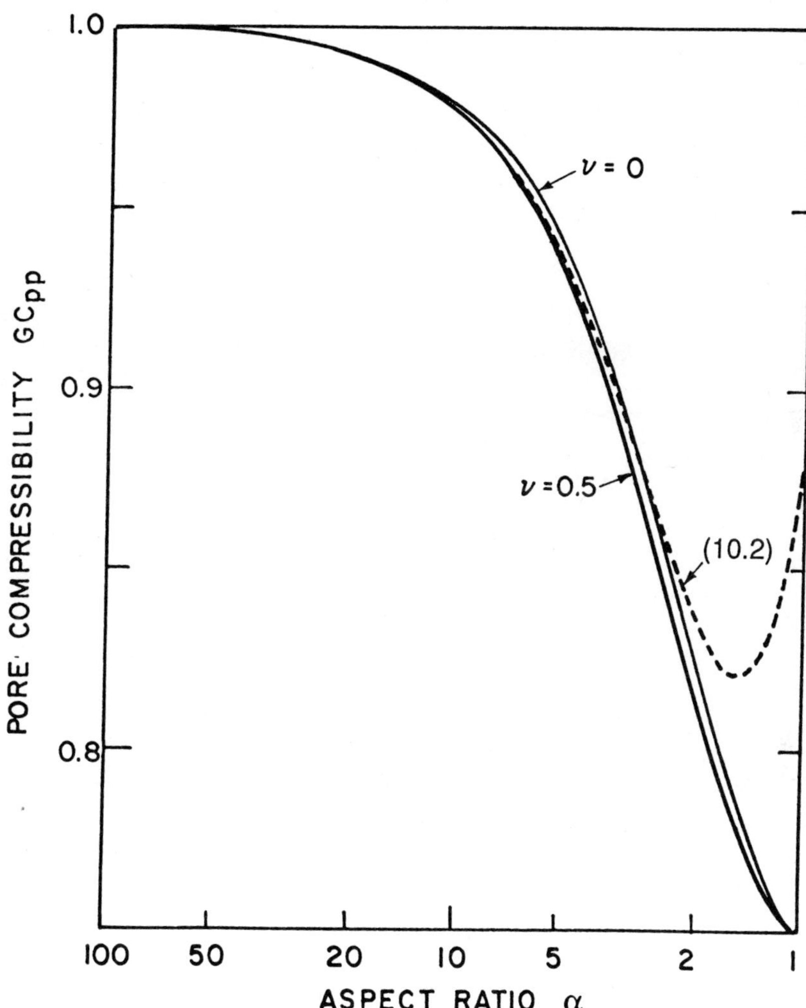

Fig. 10.3. Pore compressibility of a prolate spheroidal pore, according to the exact equation (10.1), and the asymptotic expression (10.2).

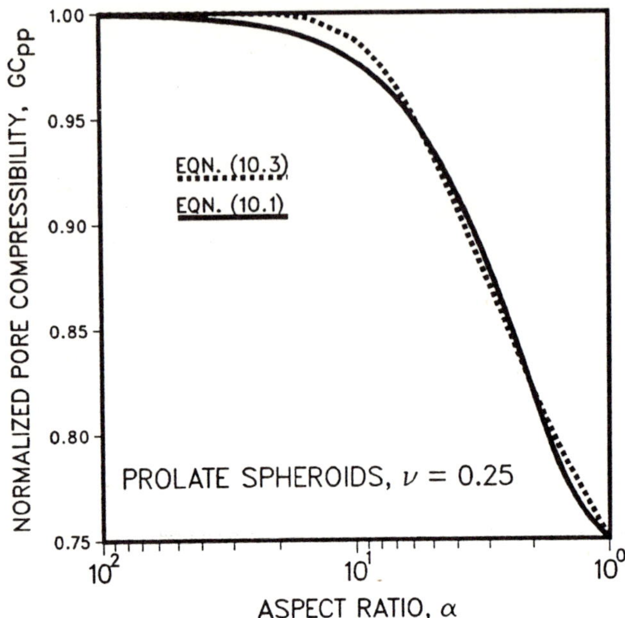

Fig. 10.4. Pore compressibility of a prolate spheroidal pore, according to the exact equation (10.1), and the approximate equation (10.3).

The slope of the C_{pp} *vs.* α curve is zero in the vicinity of $\alpha = 1$, and grows like $1/\alpha$ as $\alpha \to 0$. An asymptotic expression for the compressibility of very thin penny-shaped cracks can be found by starting with the first two terms of the Taylor series expansion of R around $\alpha = 0$; extensive algebraic manipulation then leads to (see Fig. 10.5)

$$C_{pc} = \frac{2(1-\nu)}{\pi \alpha G} + \text{terms of order } (\alpha) \quad \text{as } \alpha \to 0. \tag{10.5}$$

Strictly speaking, this result follows from the theory of linear elasticity, and does not account for the change in geometry that accompanies crack closure. However, as was the case for two-dimensional elliptical cracks, more careful analysis [Sneddon, 1946] shows that the shape of the cavity remains spheroidal as it closes, up until the point at which the two faces meet. Hence eqn. (10.5), although derived within the framework of small-strain linear elasticity, actually holds during the entire closure process. Expression (10.5) also agrees with the result that can be inferred from the work of Sack [1946], who let $\alpha \to 0$ *before* solving the equations of elasticity.

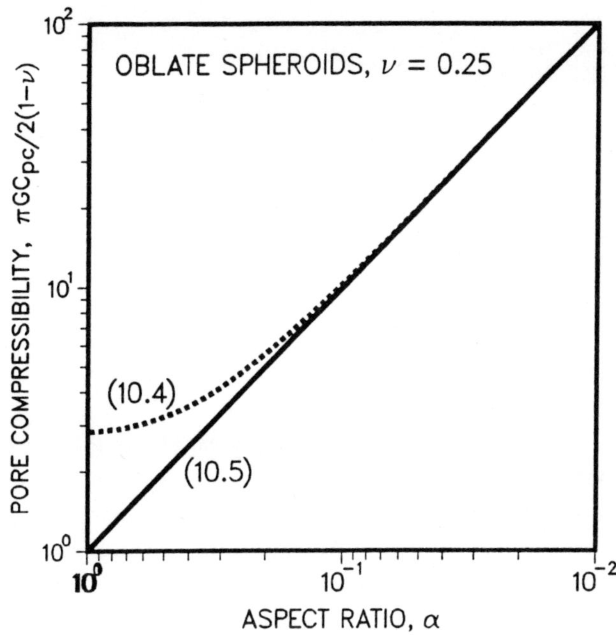

Fig. 10.5. Pore compressibility of an oblate spheroidal pore, according to the exact equation (10.4), and the asymptotic expression (10.5).

As is the case for the two-dimensional elliptical crack, C_{pc} is proportional to $(1-\nu)/\alpha G$; the constants of proportionality differ by only a factor of $\pi/2$ [Walsh, 1965a]. The closing pressure of a penny-shaped crack is $\pi\alpha G/2(1-\nu)$, which is 57% greater than that of a two-dimensional elliptical crack. A surprising feature of the asymptotic result (10.5) is its range of accuracy, which extends far beyond what is usually thought of as a "thin" crack (see Fig. 10.5). For example, when $\nu = 0.25$, the asymptotic result is accurate to within 10% for values of α as high as 0.20. The fact that eqn. (10.5) can be used over almost the entire range of aspect ratios for oblate spheroids is very useful when attempting to infer aspect ratio distributions from stress-strain curves. Methods for performing these inversions are discussed in Chapter 12.

Chapter 11. Effective Elastic Moduli

The discussion of the compressibility of pores of different shapes in Chapters 8-10 was restricted to calculations based on the idealization that a given pore was surrounded by an infinite expanse of intact rock. Pore compressibility was found by calculating the fractional volume change of a single cavity that is subjected to a hydrostatic pressure over its surface. The total pore volume change of the entire pore space was implicitly assumed to merely be the sum of the volume changes calculated for the individual isolated pores. In an actual porous rock, the stress field that a given pore is subjected to will be influenced by the presence of nearby pores, and so its volumetric strain will be different than that which would be calculated by ignoring the other pores. The effect of neighboring pores can easily be seen to increase the pore compressibility of a given pore, by the following argument. Imagine a body permeated with various pores, only one of which is pressurized by the pore fluid. The expansion of this pore will be resisted not by an infinite expanse of non-porous matrix material, but by the porous material, which has lower elastic moduli. Hence it is plausible that the pressurized pore will expand by a greater amount than it would if it were actually imbedded in the hypothetical non-porous matrix that was assumed for the calculations in Chapters 8-10. This effect is often referred to by the term ''pore interactions'', but its should be understood that this term refers here to the interactions of the stress fields surrounding the various pores, and not to physical interactions, such as crack coalescence, that may occur during inelastic deformation.

Since the compressibility of an isolated pore can be calculated by solving the suitable elastostatic boundary value problem, it might be thought that this same approach could be carried out in the case of multiple pores. Unfortunately, exact solutions to elasticity problems involving multiple cavities are very difficult to obtain. Some solutions exist for bodies containing a pair of pores, such as two spherical cavities [Sternberg and Sadowsky, 1952; Willis and Bullough, 1969], or two cylindrical cavities [Ling, 1948; Zimmerman, 1988]. It is also not yet clear how these two-pore solutions can be used to find the effective moduli of a body that contains many pores. Chen and Acrivos [1978] attempted to do this for spherical pores, but their extremely lengthy calculations merely lead to an expression that is accurate to second-order in porosity, as opposed to the first-order accuracy of the no-interaction results. The results of Chen and Acrivos are therefore accurate out to somewhat larger values of porosity, but still become unreliable for values of $\phi > 0.40$. There are even a few semi-analytical solutions to problems involving periodic arrays of cavities in two or three dimensions [Koiter, 1959; Delameter *et al.*, 1975; Horii and Nemat-Nasser, 1985]. Ordered, periodic arrays of pores, however, lead to highly anisotropic behavior that is very dissimilar to that of a body with randomly distributed pores. For example, a plate containing an hexagonal array of circular holes may have a Poisson's ratio in one direction that exceeds 0.50 [Fil'shtinskii, 1964], the theoretical maximum for an isotropic material.

The most useful approaches to the problem of predicting the compressibility of a body containing a finite concentration of pores are therefore those methods which somehow avoid dealing explicitly with more than one pore at a time. Numerous such methods have been proposed in the fields of rock physics, ceramics, and materials science, although many of these are variations on a few basic approaches. These methods are all applicable to the more general problem of predicting the effective elastic moduli of heterogeneous, multi-component materials [Watt *et al.*, 1976]. Such methods are typically discussed in a context in which both of the elastic moduli, for example K and G, are predicted. In fact, for some of these methods, the two elastic moduli must be found simultaneously. In contrast to the discussion of Chapters 8-10, for which it was convenient to consider the application of a pore pressure with zero "confining" stresses, these effective moduli methods utilize stresses imposed at the outer boundaries of the bodies. It is still possible, however, to relate these methods to the discussion of porous rock compressibilities in Chapter 2.

A convenient starting point for many of these methods are the following strain-energy considerations, which are due to Eshelby [1957]. If an homogeneous elastic body of volume V_b and bulk modulus K_r is subjected to a uniform hydrostatic compression of magnitude P over its entire outer surface, the amount of elastic energy Σ stored in the body will equal $P^2V_b/2K_r$. If it is now imagined that pores are introduced into this body, while maintaining the hydrostatic boundary conditions, the total elastic strain energy will change by an amount $\Delta\Sigma$(hydrostatic), to a new value $\Sigma = P^2V_b/2K_r + \Delta\Sigma$(hydrostatic). The effective bulk modulus K can now be defined so that the strain energy is still related to the pressure and volume in the usual way:

$$\frac{P^2V_b}{2K} = \Sigma = \frac{P^2V_b}{2K_r} + \Delta\Sigma(\text{hydrostatic}) . \tag{11.1}$$

Similar considerations, when the boundary conditions would cause a state of pure shear of magnitude T to exist in the non-porous body, lead to

$$\frac{T^2V_b}{2G} = \frac{T^2V_b}{2G_r} + \Delta\Sigma(\text{shear}) . \tag{11.2}$$

Eqns. (11.1) and (11.2) serve to define the effective elastic moduli $\{K,G\}$. Another reasonable definition of the effective elastic moduli of a heterogeneous body would be the ratios of the average stresses to the average strains that result in the body when it is subject to pure shear or pure compression on its outer boundary. Although it is not easy to prove [see Hashin, 1983], this definition is equivalent to the energy-based definition proposed by Eshelby, in that both lead to the same values of the effective moduli.

It is possible to relate the above energy considerations to the porous rock compressibilities defined in Chapter 1. The energy perturbation $\Delta\Sigma$(hydrostatic) can be thought of as being composed of three terms. First, imagine that the pores are "carved out" of the body, while somehow the appropriate stress are maintained at the pore boundary so as not to allow the pore to relax. The body will thereby lose an amount of strain energy equal to that which was stored in the carved-out material, *i.e.*, $P^2V_p/2K_r = C_rP^2V_p/2$. Now imagine that these stresses at the pore boundary are relaxed slowly to zero. As the pore surface relaxes, its volume will change by an amount equal to $-C_{pp}V_pP$, and the body will perform an amount of work against this stress that is equal to $C_{pp}V_pP^2/2$. This term therefore also represents energy lost by the body. Finally, as the fictitious pore stresses are relaxed, the outer boundary of the body will contract by an amount equal to $-C_{bp}V_bP$. Since the pressure on the outer boundary remains unchanged at P during this process, the external stress will perform an amount of work on the body that is equal to $C_{bp}V_pP^2$. The total energy perturbation is therefore given by

$$\Delta\Sigma(\text{hydrostatic}) = -(P^2V_PC_r/2) - (P^2V_pC_{pp}/2) + (P^2V_bC_{bp})\,. \tag{11.3}$$

If eqns. (2.2) and (2.5) are used to express C_{bp} and C_{pp} in terms of C_{pc}, it is seen that $\Delta\Sigma(\text{hydrostatic}) = P^2\phi V_bC_{pc}/2$. This result, combined with eqn. (11.1) and the fact that $C_{bc} = 1/K$, leads back to the relationship $C_{bc} = C_r + \phi C_{pc}$. The strain energy perturbation due to the addition of a pore in a body under hydrostatic compression is therefore exactly equal to $P^2\phi V_bC_{pc}/2$.

Utilization of eqns. (11.1) and (11.2) to define the effective moduli require the estimation of the energy perturbation terms for hydrostatic and shear loading. The simplest approach is to consider the introduction of one typical pore into the body, calculate the resulting energy change, and then to sum up this energy for all of the pores. If the single pores are assumed to be very small relative to the overall dimensions of the body, this method is equivalent to the approach taken in Chapters 8-10. The two energy expressions, $\Delta\Sigma$(hydrostatic) and $\Delta\Sigma$(shear), are then functions of $\{K_r, G_r\}$, and are linearly proportional to the porosity, ϕ. Eqns. (11.1) and (11.2) are thus uncoupled, and can be independently solved for the effective elastic moduli $\{K,G\}$. This approach ignores stress field interactions between pores, and, as implied earlier, usually underestimates the effective elastic compliances, thus overestimating the effective moduli.

A method of accounting for strain-energy interactions of the different pores, known as the "self-consistent" scheme, was proposed independently by Hill [1965] and Budiansky [1965]. This method calculates the strain energy due to a typical pore by assuming that it is introduced into an homogeneous medium which has the elastic properties of the actual porous material. This leads to the same functional forms for the two energy terms as does the "no-interaction" approach, except that $\{K_r, G_r\}$ are replaced by $\{K, G\}$ as the arguments of these functions. Eqns. (11.1) and (11.2) then take the form of two coupled algebraic equations for the effective moduli. For some pore shapes, these equations can be solved in closed form for $\{K, G\}$, while for others the inversion must be carried out numerically.

It has been suggested by Bruner [1976] that the self-consistent method overestimates the effect that pores have in lowering the elastic moduli, by implicitly taking interactions between pairs of pores into account twice. A modification of the self-consistent approach which presumably avoids this error has been outlined by Salganik [1973] and Henyey and Pomphrey [1982] for the specific case of penny-shaped cracks, and by Zimmerman [1984b] for spherical pores. In this modified self-consistent method, the pores are introduced into the body one at a time, with the $(n+1)st$ pore considered to be placed into an homogeneous medium which has the effective elastic properties of the body with n pores. In this way, the $(n+1)st$ pore feels the effect, so to speak, of the nth pore, but not *vice versa,* avoiding the error of counting this energy interaction twice. This process is illustrated schematically in Fig 11.1. We begin with a homogeneous, non-porous body whose elastic moduli are represented by M_r. Next a small amount of porosity, $\Delta\phi$, is introduced into the material, after which the new effective moduli are calculated. Since $\Delta\phi$ is small, these new moduli can be found "exactly" through the non-interactive method. We now replace this porous material with an equivalent homogeneous body whose moduli are $M(\Delta\phi)$, after which the process is repeated. The energy terms in eqns. (11.1) and (11.2) are then case functions of the "instantaneous" moduli $\{K(\phi), G(\phi)\}$, and so in the limit as $\Delta\phi \to 0$, these equations take the form of two coupled, nonlinear ordinary differential equations.

To be more specific, consider the case of spherical pores. For the addition of pores which contribute an additional incremental porosity $\Delta\phi$, the energy terms take the following forms [Zimmerman, 1984b]:

$$\Delta\Sigma(\text{hydrostatic}) = \frac{P^2 V_b}{2K}\left[1 + \frac{3K}{4G}\right]\Delta\phi\ , \tag{11.4}$$

$$\Delta\Sigma(\text{shear}) = \frac{T^2 V_b}{2G}\left[\frac{15K + 20G}{9K + 8G}\right]\Delta\phi\ . \tag{11.5}$$

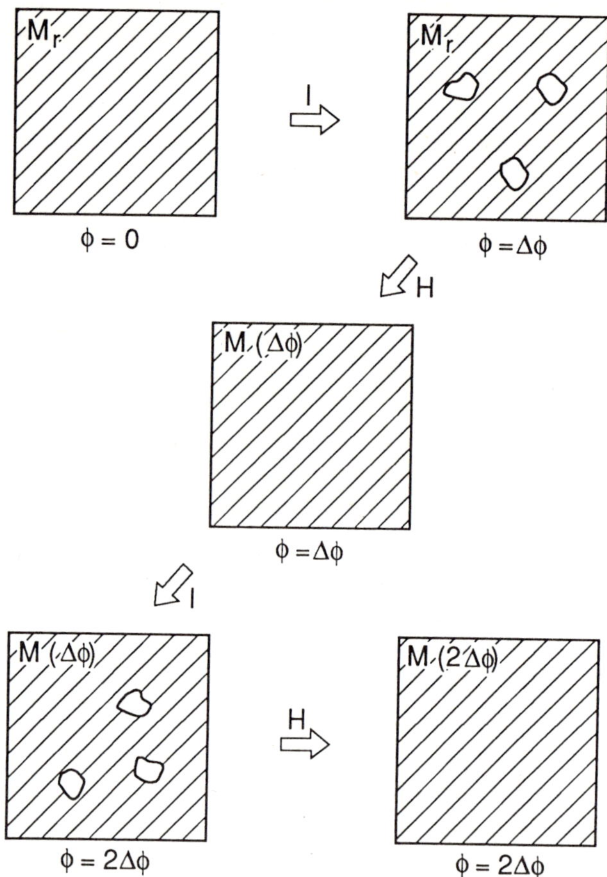

Fig. 11.1. Schematic diagram illustrating the differential scheme for estimating effective elastic moduli. In each step of the process, small amounts of porosity, $\Delta\phi$, are imagined to be introduced into the body, after which the new moduli are calculated by the "no-interaction" method. In the limit as $\Delta\phi \to 0$, the result is a pair of coupled ordinary differential equations that describe the evolution of the moduli with ϕ.

For the no-interaction method, where the moduli to be used in the energy expressions are $\{K_r, G_r\}$ regardless of how much porosity has been introduced, eqns. (11.1) and (11.2) can readily be solved for

$$\frac{K_r}{K(\phi)} = 1 + \left[1 + \frac{3K_r}{4G_r}\right]\phi\,, \tag{11.6}$$

$$\frac{G_r}{G(\phi)} = 1 + \left[\frac{15K_r + 20G_r}{9K_r + 8G_r}\right]\phi\,. \tag{11.7}$$

For the self-consistent method, if the introduction of a *finite* amount of porosity ϕ is considered, the arguments of the energy terms in eqns. (11.4) and (11.5) are the moduli $\{K(\phi), G(\phi)\}$, whereafter eqns. (11.1) and (11.2) take the forms

$$\frac{1}{K(\phi)} = \frac{1}{K_r} + \left[\frac{1}{K(\phi)} + \frac{3}{4G(\phi)}\right]\phi\,, \tag{11.8}$$

$$\frac{1}{G(\phi)} = \frac{1}{G_r} + \left[\frac{15K(\phi) + 20G(\phi)}{9K(\phi) + 8G(\phi)}\right]\frac{\phi}{G(\phi)}\,. \tag{11.9}$$

For most values of the Poisson ratio of the matrix phase, these equations must be solved numerically, since they are algebraically nonlinear in $K(\phi)$ and $G(\phi)$.

For the differential self-consistent method, infinitesimal changes in the porosity must be considered, in which case eqns. (11.1) and (11.2) take the form

$$\frac{1}{K(\phi+d\phi)} = \frac{1}{K(\phi)} + \frac{1}{K(\phi)}\left[1 + \frac{3K(\phi)}{4G(\phi)}\right]d\phi\,, \tag{11.10}$$

$$\frac{1}{G(\phi+d\phi)} = \frac{1}{G(\phi)} + \frac{1}{G(\phi)}\left[\frac{15K(\phi)+20G(\phi)}{9K(\phi)+8G(\phi)}\right]d\phi\,. \tag{11.11}$$

Forming the Newton quotients $[K(\phi+d\phi)-K(\phi)]/K(\phi)$ and $[G(\phi+d\phi)-G(\phi)]/G(\phi)$, and passing to the limit of $d\phi \to 0$, the following differential equations are arrived at:

$$-\frac{1}{K}\frac{dK}{d\phi} = 1 + \frac{3K}{4G}\,, \tag{11.12}$$

$$-\frac{1}{G}\frac{dG}{d\phi} = \frac{15K+20G}{9K+8G}\,. \tag{11.13}$$

The moduli of the matrix material $\{K_r, G_r\}$ appear now only as initial conditions on these equations:

$$K(\phi=0) = K_r \ , \tag{11.14}$$

$$G(\phi=0) = G_r \ . \tag{11.15}$$

The solution to these equations is [Norris, 1985]:

$$\frac{G}{G_r} = e^{-2\phi}\left[\frac{1.5 + \lambda(G/G_r)^{3/5}}{1.5+\lambda}\right]^{1/3} , \tag{11.16}$$

$$\frac{K}{K_r} = \frac{G}{G_r}\left[\frac{0.75 + \lambda}{0.75 + \lambda(G/G_r)^{3/5}}\right] , \tag{11.17}$$

where $\lambda = 3(1-5\nu_r)/4(1+\nu_r)$. Although these expressions are implicit, and need to be solved numerically to yield the moduli, two-digit accuracy can be obtained by first substituting $G/G_r = e^{-2\phi}$ into the right-hand side of eqn. (11.16), and then solving eqns. (11.16) and (11.17) in turn.

Norris [1985] has provided a slightly different derivation of the differential self-consistent equations than that given by Zimmerman [1984b]. Zimmerman assumed that when a new pore is introduced into the body, it occupies space previously taken up by solid material. Norris, on the other hand, assumes that each new pore may displace either matrix material, or a pore, with probabilities $1-\phi$ and ϕ, respectively. This method leads to the same differential equations (11.12) and (11.13), except that $d\phi$ is replaced by $d\phi/(1-\phi)$. This different manner of parametrizing the porosity increments has the effect of replacing $e^{-2\phi}$ in eq. (11.16) with $(1-\phi)^2$. Note that this process of placing pores in the body is nothing more than a "thought experiment" that allows a calculation of the effective moduli, and is not intended to model the actual evolution of porosity in a sandstone.

The final method of predicting effective elastic moduli that will be discussed here is not based on the energy considerations expressed by eqns. (11.1) and (11.2). Kuster and Toksöz [1974] calculated the resultant of the elastic waves which have been scattered once by each of an assemblage of pores (Fig. 11.2) in a body with moduli $\{K_r, G_r\}$, and equated this to the wave which would be scattered by an "equivalent homogeneous spherical inclusion" whose moduli are $\{K, G\}$, in order to arrive at expressions for the velocities of transverse (shear) and longitudinal (compressional) waves, V_T and V_L. Although this method is based on the

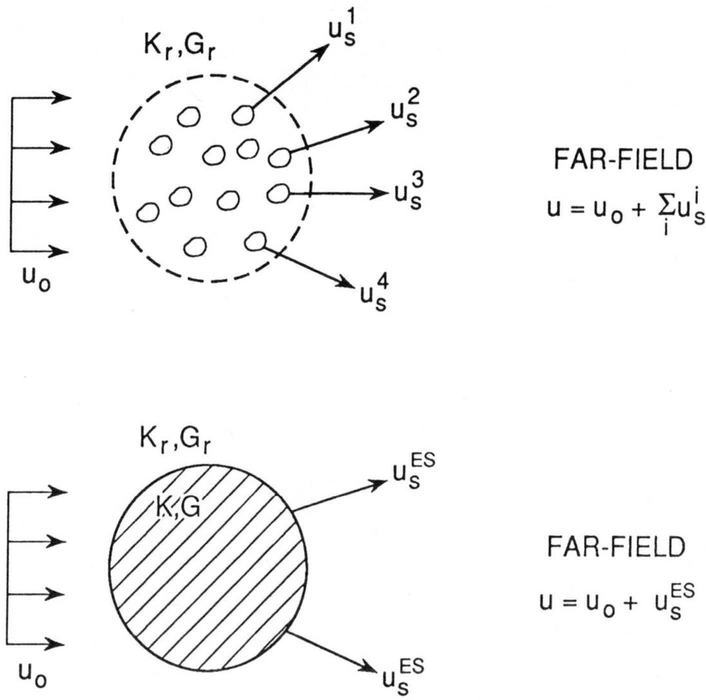

Fig. 11.2. Schematic diagram illustrating the Kuster-Toksöz [1974] method for estimating effective elastic moduli. The elastic wave that is scattered (once, without considering multiple scattering) from an assemblage of pores in a body with moduli $\{K_r, G_r\}$ is compared with the wave that would be scattered by a spherical inclusion of *effective moduli* $\{K,G\}$. Equating the amplitudes of these two waves leads to an expression for the effective moduli.

dynamic process of wave-propagation, in contrast to the static energy considerations upon which the other methods are based, the results can be converted into effective elastic moduli through the relationships [Sokolnikoff, 1956] $G = \rho V_T^2$ and $K = \rho[V_L^2 - 4V_T^2/3]$, where ρ is the average density. For a porous rock with vacuous pores, $\rho = (1-\phi)\rho_r$. (Recall that K and G are implicitly the *drained* moduli, since the pores are assumed to be evacuated. The moduli appropriate for wave propagation are, however, the *undrained* moduli. Although G is the same for both types of process, $K = K_u$ only when the pores are "dry".) Multiple scattering of the waves was not considered, so that this approach does not explicitly take into account the interactions between pores. Despite this seeming limitation, the Kuster-Toksöz approach cannot be dismissed, and is certainly more accurate than the "no-interaction" method. In fact, this method is perhaps the effective moduli theory that has been most widely used by geophysicists. For example, Toksöz *et al.* [1976] and Wilkens *et al.* [1986] used this theory to model seismic velocities in reservoir sandstones, and Zimmerman and King [1986] used the theory to study the effect of the ice/water ratio on seismic velocities in permafrost. For a body containing dry spherical pores, the Kuster-Toksöz method predicts

$$\frac{K(\phi)}{K_r} = \frac{1-\phi}{1 + \left[\dfrac{3K_r}{4G_r}\right]\phi} , \tag{11.18}$$

$$\frac{G(\phi)}{G_r} = \frac{1-\phi}{1+\left[\dfrac{6K_r+12G_r}{9K_r+8G_r}\right]\phi} . \tag{11.19}$$

As an example of how these five approaches differ in their predictions, consider the case where $G_r = 0.75K_r$, which corresponds to a matrix Poisson ratio of 0.20. In this particular case it is possible to obtain simple, explicit solutions for the effective moduli for all five theories. The no-interaction theory (eqns. (11.6 and 11.7)) and the Kuster-Toksöz theory (eqns. (11.18) and (11.19)) always lead to simple explicit solutions for the moduli. Since the λ parameter that appears in eqns. (11.16) and (11.17) vanishes when $\nu_r = 0.20$, the equations of the differential scheme simplify considerably in this case. An explicit solution to the equations of the self-consistent scheme also exists when $\nu_r = 0.20$, although it is difficult to see this from eqns. (11.8) and (11.9). This value of ν_r is also unique in that it always leads to the same expression for $K(\phi)/K_r$ as it does for $G(\phi)/G_r$. The solutions for $\nu_r = 0.20$ are

no–interaction: $$\frac{K(\phi)}{K_r} = \frac{G(\phi)}{G_r} = \frac{1}{1+2\phi} ; \tag{11.20}$$

self–consistent: $$\frac{K(\phi)}{K_r} = \frac{G(\phi)}{G_r} = 1-2\phi ; \tag{11.21}$$

differential (Zimmerman): $$\frac{K(\phi)}{K_r} = \frac{G(\phi)}{G_r} = e^{-2\phi} ; \tag{11.22}$$

differential (Norris): $$\frac{K(\phi)}{K_r} = \frac{G(\phi)}{G_r} = (1-\phi)^2 ; \tag{11.23}$$

Kuster–Toksöz: $$\frac{K(\phi)}{K_r} = \frac{G(\phi)}{G_r} = \frac{1-\phi}{1+\phi} . \tag{11.24}$$

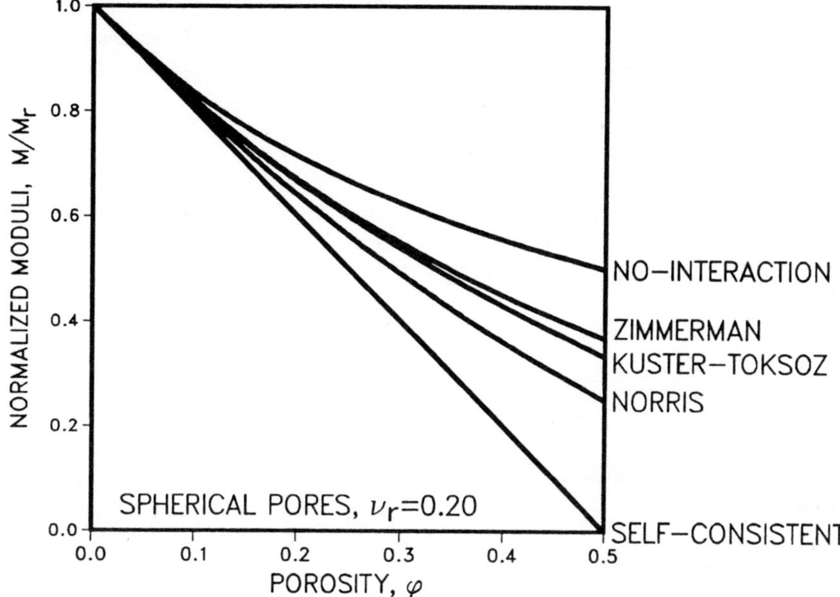

Fig. 11.3. Normalized elastic moduli of a solid containing spherical pores, according to eqns. (11.20-11.24). The Poisson ratio of the matrix material is taken to be 0.20.

This particular value of $\nu_r = 0.20$ lies within the range commonly found for rock-forming minerals, yet at the same time the predictions of the theories are qualitatively similar for other values of ν_r.

The normalized moduli predicted by these five theories are plotted in Fig. 11.3, for porosities up to 50%. While each of the four theories predicts the same first-order in ϕ term (*i.e.*, $1-2\phi+\ldots$), the results begin to diverge noticeably for porosities as low as 10%. The Kuster-Toksöz results for spherical pores coincide with the Hashin-Shtrikman [1961] upper bounds, so it is clear that the no-interaction method will substantially overestimate the moduli at moderate porosities. The self-consistent method predicts that the moduli go to zero for porosities above 50%, which seems to indicate that this method will not be very accurate at moderate or high porosities. Zimmerman's version of the differential method predicts moduli that slightly exceed the Hashin-Shtrikman upper bounds, although the error does not become pronounced until the porosity reaches about 50%. Norris's version of the differential method falls below the upper bound, and also has the advantage of predicting that the moduli go to zero as $\phi \rightarrow 100\%$.

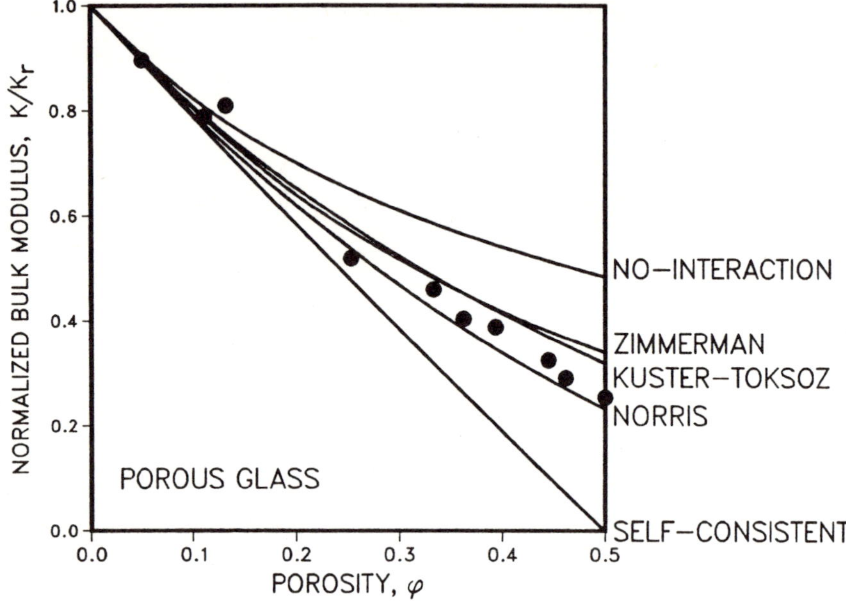

Fig. 11.4. Normalized bulk modulus of a porous glass containing nearly-spherical pores. The Poisson ratio of the matrix is 0.23; data are from Walsh *et al.* [1965].

For for the range of porosities commonly encountered in sandstones, the various theories are not in close agreement, at least not when applied to materials with spherical pores (Fig. 11.3). Although there have been many heuristic arguments put forward in favor of each of the effective moduli theories, the most unambiguous way to discern the relative accuracies of the various methods would be to compare their predictions with experimental data. Since the pores in sandstones are of course never exactly spherical or penny-shaped, comparison with real compressibility data would not reveal whether any discrepancies lie in the model equations, or in the non-idealized shape of the pores; it would be more instructive to test the theories against materials which are known to have spherical pores. There unfortunately is not much existing data on the elastic moduli of materials with spherical pores, at least not at porosities high enough to allow for discrimination between the different theories. Two sets of applicable data are the bulk modulus measurements of Walsh *et al.* [1965] on a porous glass with $\nu_r = 0.23$, and the measurements of the compressional wave modulus $M = K + 4G/3$ made by Warren [1969] on a sintered perlite with $\nu_r = 0.193$. In each case, the materials were specially fabricated with spherical pores for the purposes of testing various effective moduli theories. For the latter set of data, only specimens with porosities below about 20% had pores which were spherically shaped. These data are plotted in Figs. 11.4 and 11.5, respectively, along with the

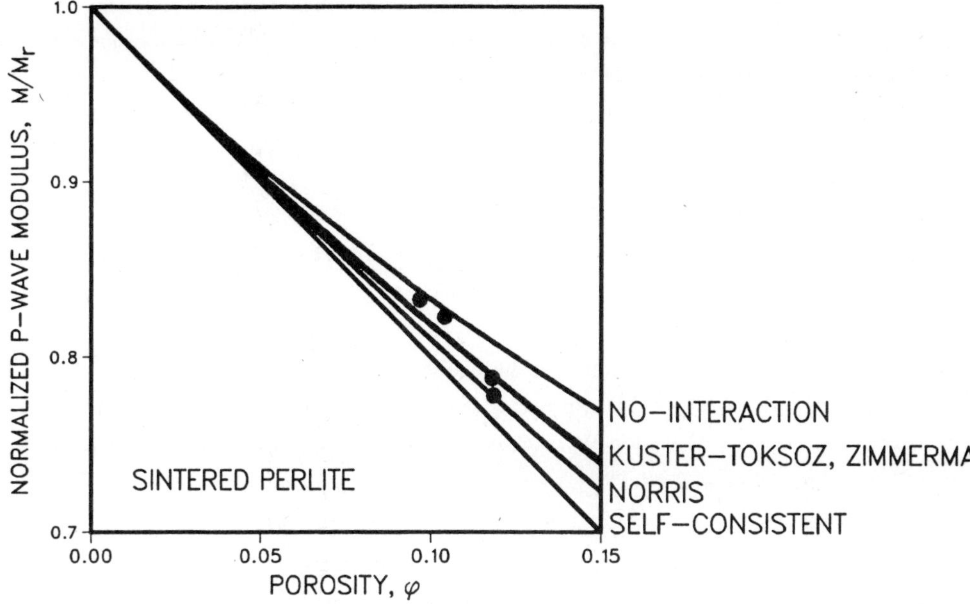

Fig. 11.5. Normalized P-wave modulus ($M = K + 4G/3$) of a sintered perlite containing spherical pores. The Poisson ratio of the matrix is 0.193; the data are from Warren [1969].

predictions of the five theories: the no-interaction theory, the self-consistent theory, the two versions of the differential theory, and the Kuster-Toksöz theory. For both sets, the data straddle the curves predicted by the differential schemes and and the Kuster-Toksöz theory for low porosities. The Kuster-Toksöz theory provides the best fit to these sets of data, followed in accuracy by Norris's differential scheme. The complete lack of coherence predicted by the original self-consistent theory does not occur, and in fact there is one datum point for a sample of glass with 70% porosity [Walsh *et al.*, 1965] that not only has a nonzero bulk modulus, but which lies only 20% below the Kuster-Toksöz predictions.

Each of the five methods described above can, in principle, be used to predict the elastic moduli of solids containing pores of any shape. One of the most important shapes for rock mechanics applications is the thin oblate-spheroidal penny-shaped crack. All that is needed for the application of the various effective moduli theories to this pore shape are the two energy expressions involved in eqns. (11.1) and (11.2), which have been found by Budiansky and O'Connell [1976]. Before presenting the predictions of the various theories for the case of penny-shaped cracks, however, some discussion of the terminology used for this problem is needed. As is seen from eqn. (10.5), the pore compressibility of an isolated penny-shaped crack is inversely proportional to the aspect ratio α. Eqn. (11.3) shows that the energy associated with a pressurized crack is proportional to the *product* of its compressibility and its volume. Now consider a penny-shaped crack of aspect ratio α whose semi-major axis is of length r, and whose semi-minor axis is of length αr. The volume of this particular crack will be $4\pi\alpha r^3/3$. The product of the volume of the crack and its compressibility does not involve α, since it cancels out in the numerator (from the volume term) and in the denominator (from the compressibility term). The important parameter describing the effect of a crack in an elastic medium is therefore not the volume of the crack, but rather the parameter r^3. For more than one crack, the important parameter is $\Gamma = Nr^3/V_b$, where N/V_b is the number of cracks per unit bulk volume. Following Budiansky and O'Connell, this parameter is usually referred to as the "crack density parameter".

In order to apply the effective moduli theories to a body with penny-shaped cracks, the strain energy perturbations $\Delta\Sigma$(hydrostatic) and $\Delta\Sigma$(shear) are needed. The hydrostatic strain energy perturbation $\Delta\Sigma$(hydrostatic) caused by an isolated penny-shaped crack is equal to $P^2\phi V_b C_{pc}/2$, with C_{pc} given by eqn. (10.5):

$$C_{pc} = \frac{2(1-\nu)}{\pi G \alpha} = \frac{4(1-\nu^2)C}{3\pi(1-2\nu)\alpha} \; . \tag{11.25}$$

The porosity represented by a single crack of radius r and aspect ratio α is $4\pi r^3\alpha/3V_b$, so the porosity represented by N such cracks is $4N\pi r^3\alpha/3V_b$. Hence

$$\Delta\Sigma(\text{hydrostatic}) = \frac{P^2\phi V_b C_{pc}}{2}$$

$$= \frac{P^2V_b}{2}\left[\frac{4N\pi r^3\alpha}{3V_b}\right]\left[\frac{4(1-\nu^2)C}{3\pi(1-2\nu)\alpha}\right]$$

$$= \frac{8(1-\nu^2)\Gamma}{9(1-2\nu)} \frac{P^2 V_b}{K} \; . \tag{11.26}$$

Eqn. (11.26) confirms that Γ, and not ϕ, is the parameter that influences the effective compressibility of the rock. To within a high degree of approximation, the aperture of a crack has no explicit effect on the moduli. The aperture of a crack does, of course, determine the pressure at which the crack closes, and thus influences the nonlinear shape of the stress-strain curve (see Chapter 12).

To find the strain energy perturbation $\Delta\Sigma$(shear), it is first necessary to find the strain energy of an isolated crack subjected to a shear stress of magnitude T at infinity. The resulting strain energy expression will include a dependence on the angle between the normal vector of the crack and the applied load. If the cracks are randomly oriented in the rock, this energy must then be averaged over all possible angles of inclination. (In principle, of course, the cracks may have preferred orientations, in which case weighting functions would be needed when carrying out the averaging process. Nishizawa [1982], for example, used the differential scheme to predict the elastic moduli of an anisotropically cracked rock.) For an isotropic distribution of cracks, Budiansky and O'Connell [1976] performed the required averaging, and found that

$$\Delta\Sigma\,(\text{shear}) = \frac{16(1-\nu)(5-\nu)\Gamma}{45(2-\nu)}\,\frac{T^2 V_b}{G}. \tag{11.27}$$

According to the no-interaction scheme, the effective elastic moduli of an isotropically cracked body can be found by combining the general expressions (11.1) and (11.2) with the energy perturbations (11.26) and (11.27), with $\{K_r, \nu_r\}$ used in the expressions for Σ:

$$\frac{K_r}{K} = 1 + \frac{16(1-\nu_r^2)\Gamma}{9(1-2\nu_r)}\,, \tag{11.28}$$

$$\frac{G_r}{G} = 1 + \frac{32(1-\nu_r)(5-\nu_r)\Gamma}{45(2-\nu_r)}\,. \tag{11.29}$$

Since the effect of each crack is assumed to be additive, the compliances $1/K$ and $1/G$ are linear functions of the crack density parameter Γ.

The predictions of the self-consistent method are also found by combining eqns. (11.1) and (11.26), and eqns. (11.2) and (11.27), with $\{K, \nu\}$ used in the expressions for Σ. This leads to the following implicit expressions for $\{K, G\}$ [Henyey and Pomphrey, 1982]:

$$\frac{K}{K_r} = 1 - \frac{16(1-\nu^2)\Gamma}{9(1-2\nu)}\,, \tag{11.30}$$

$$\frac{G}{G_r} = 1 - \frac{32(1-\nu)(5-\nu)\Gamma}{45(2-\nu)}\,, \tag{11.31}$$

$$\Gamma = \frac{45(\nu_r - \nu)(2-\nu)}{16(1-\nu^2)(10\nu_r - 3\nu\nu_r - \nu)}\,, \tag{11.32}$$

where $\nu=(3K-2G)/(6K+2G)$. To calculate the effective moduli for a given value of Γ, one would first solve eqn. (11.32) for ν, and then find $\{K,G\}$ from eqns. (11.30) and (11.31). Since eqn. (11.32) is implicit in the unknown Poisson ratio ν, it is inconvenient. Although no explicit solution exists for the moduli in terms of Γ, the following approximate expression for ν as a function of Γ is extremely accurate [O'Connell and Budiansky, 1974]:

$$\frac{\nu}{\nu_r} = 1 - \frac{16\Gamma}{9} . \tag{11.33}$$

As mentioned above, the two versions of the differential scheme are equivalent, except for the manner in which the porosity is parametrized. The equations of Zimmerman's version can be transformed into those of Norris's version by replacing ϕ with $-\ln(1-\phi)$. For cracks, the porosity is very small, and the two parametrizations are equivalent, since $-\ln(1-\phi)=\phi$, to first order. A rigorous discussion of the application of the differential scheme to cracked bodies has been given by Hashin [1988]. The differential equations for the elastic moduli are

$$-\frac{1}{K}\frac{d\ln K}{d\Gamma} = \frac{16(1-\nu^2)}{9(1-2\nu)} , \tag{11.34}$$

$$-\frac{1}{G}\frac{d\ln G}{d\Gamma} = \frac{32(1-\nu)(5-\nu)}{45(2-\nu)} . \tag{11.35}$$

An implicit solution to eqns. (11.34) and (11.35) was found by Zimmerman [1985c]:

$$\frac{128\Gamma}{5} = \ln\left[\frac{3-\nu}{3-\nu_r}\right] + 6\ln\left[\frac{1-\nu}{1-\nu_r}\right] + 9\ln\left[\frac{1+\nu}{1+\nu_r}\right] - 16\ln\left[\frac{\nu}{\nu_r}\right] , \tag{11.36}$$

$$\frac{K}{K_r} = \left[\frac{\nu}{\nu_r}\right]^{10/9}\left[\frac{3-\nu}{3-\nu_r}\right]^{-1/9}\left[\frac{1-2\nu}{1-2\nu_r}\right]^{-1} . \tag{11.37}$$

A useful approximate solution to eqns. (11.34) and (11.35) was found by Bruner [1976], who noticed that if differential equations are developed for the moduli E and ν, instead of for K and G, the right-hand sides of these equations are nearly constant. By approximating the right-hand sides of these differential equations by their values at $\nu=0$, the following simple expressions were found:

$$\frac{E}{E_r} = e^{-16\Gamma/9} , \tag{11.38}$$

$$\frac{\nu}{\nu_r} = e^{-8\Gamma/5} , \tag{11.39}$$

which can be shown to correspond to the leading terms of the exact solutions.

Kuster and Toksöz [1974] presented equations for the effective moduli of a body containing spheroidal pores of arbitrary aspect ratio, but did not explicitly consider the limiting case of infinitely thin cracks. However, if their expressions are examined in the limit as the aspect ratio goes to zero, the predicted effective moduli are found to be

$$\frac{K}{K_r} = \frac{1 - \frac{32}{27}(1+\nu_r)\Gamma}{1 + \frac{16}{27}\frac{(1+\nu_r)^2}{(1-2\nu_r)}\Gamma} , \tag{11.40}$$

$$\frac{G}{G_r} = \frac{1 - \frac{32(5-\nu_r)(37-35\nu_r)}{2025(2-\nu_r)}\Gamma}{1 + \frac{64(5-\nu_r)(4-5\nu_r)}{2025(2-\nu_r)}\Gamma} . \tag{11.41}$$

The predicted bulk moduli for a rock containing penny-shaped cracks is shown in Fig. 11.6, as a function of the crack density parameter Γ. The matrix material is taken to have a Poisson's ratio of 0.25. An important fact about this plot is that the relative positions of the curves are the same as in the case of spherical pores (Fig. 11.3). The no-interaction method predicts the highest bulk modulus (lowest compressibility), followed by the differential method, the Kuster-Toksöz method, and the self-consistent method. It seems plausible to generalize from these two cases that this ordering will hold as a rule if these theories are applied to other pore geometries, although this is not known with certainty. Hashin [1988] applied the differential self-consistent scheme and the no-interaction method to a two-dimensional body with aligned cracks, and the results do in fact show that the differential self-consistent theory predicts lower moduli than does the no-interaction method, although the differences are not nearly as large as in the case of the randomly distributed cracks shown in Fig. 11.6.

Both the self-consistent and Kuster-Toksöz theories predict that the elastic moduli vanish at some finite value of the crack density. For the case $\nu_r = 0.25$, the self-consistent theory predicts that the bulk modulus will vanish at $\Gamma = 0.5625$, and the Kuster-Toksöz theory predicts that the bulk modulus vanishes at $\Gamma = 0.675$. The situation is qualitatively similar for the predictions of the shear modulus (Fig. 11.7). The self-consistent model predicts that the shear modulus will vanish at a crack density of $\Gamma = 0.5625$, which is the same critical crack density as for the bulk modulus. It is also worth noting that this predicted critical crack density is

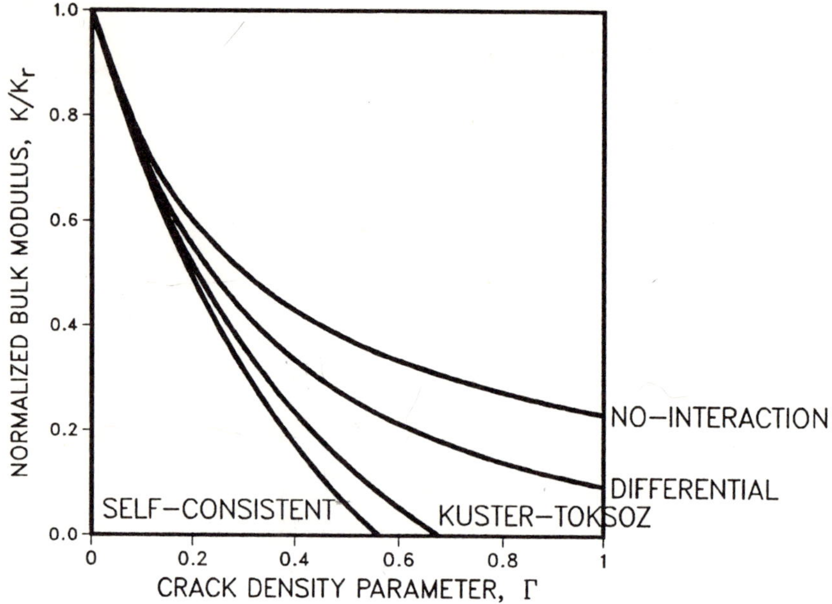

Fig. 11.6. Normalized bulk modulus of a body containing a random distribution of penny-shaped cracks, according to the various effective moduli theories. The Poisson ratio of the matrix is taken to be 0.25.

independent of the matrix Poisson ratio, ν_r. The Kuster-Toksöz theory predicts a different critical crack density for the shear modulus than for the bulk modulus; furthermore, each critical value of Γ depends on ν_r. For an matrix Poisson ratio of 0.25, the Kuster-Toksöz theory predicts a critical value of $\Gamma = 0.825$.

Whether or not this predicted vanishing of the elastic moduli is an argument for or against these effective moduli theories has been a subject of much debate [Bruner, 1976; O'Connell and Budiansky, 1976]. On the one hand, it is obviously true that a sufficient density of cracks will cause a rock to become incompetent, and unable to withstand tensile stresses, for example. On the other hand, each of the theories *starts* with the implicit assumption that the cracks do not physically intersect, in which case the elastic moduli should be nonzero. Another point to consider is that it is very difficult to imagine a rock whose crack density is greater than about one. To see this, imagine a cubic close-packed array of spherical grains of radius a, with each grain boundary representing a "crack". It is clear that the radius of each crack cannot exceed a. Since each grain will have six such cracks at its boundary, each shared with one other grain, the value of Γ will be $3a^3/(2a)^3 = 3/8$. Addition of another crack within each grain would raise Γ to about 1/2, and it is difficult to imagine a crack geometry that would lead to

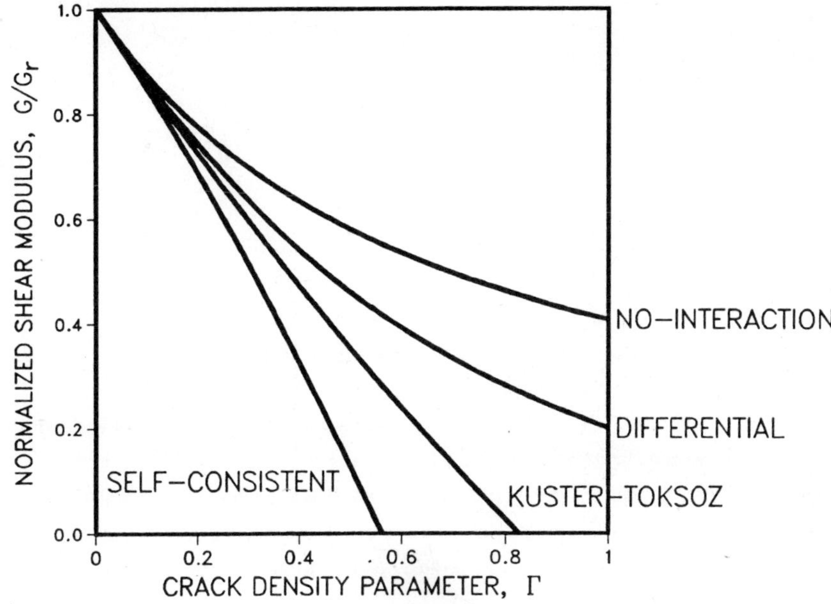

Fig. 11.7. Normalized shear modulus of a body containing a random distribution of penny-shaped cracks, according to the various effective moduli theories. The Poisson ratio of the matrix is taken to be 0.25.

much of a further increase in Γ. Although this is far from a rigorous proof that there is an upper bound on the range of feasible values of Γ, it does seem that values of $\Gamma > 1$ that may be needed to fit compressibility data to some model equations (see Chapter 12) are probably not realistic.

Chapter 12. Aspect Ratio Distributions

The preceding four chapters were devoted to the problem of relating the elastic moduli of a sandstone to the structure of its pore space. At any given level of pressure, the structure of the void space will determine the incremental elastic moduli of the rock. As the pore or confining pressure is varied, however, the void space of the rock will deform. Specifically, as the differential pressure increases, the "crack-like" voids along the boundaries between grains will close up. For example, in Chapters 9 and 10 the pressures needed to close isolated two-dimensional or three-dimensional elliptical cracks were found to be $\alpha G_r/(1-\nu_r)$, and $\pi\alpha G_r/2(1-\nu_r)$, respectively. Such closed cracks will no longer contribute to the compressibility of the rock, and will, as far as hydrostatic compression is concerned, behave as if they were welded shut. The actual crack-like voids in sandstones are of course not exactly elliptical. However, the discussion in Chapter 9 showed that for various "crack-like" void shapes, the closing pressure is always proportional to the aspect ratio. Hence as the differential pressure increases, cracks of successively higher aspect ratio will close up. The porous rock compressibilities are therefore all decreasing functions of the differential pressure, and typical stress-strain curves will be "concave-downward".

The most common approach to modeling this nonlinearity has been to assume that the rock contains a distribution of two or three-dimensional elliptical cracks of various aspect ratios. This approach has been taken by, for example, by Walsh and Decker [1966], Morlier [1971], Cheng and Toksöz [1979], Seeburger and Nur [1984], and Zimmerman [1984a]. Mavko and Nur [1978], on the other hand, pointed out that any realistic hydrostatic stress-strain curve could also be explained by assuming that all the cracks have the same, appropriate cusp-shaped profile. Some researchers, such as Siegfried and Simmons [1978], emphasizing the fact that scanning electron micrographs of cracks in rocks do not appear to be elliptical, have implied that elliptical crack models represent nothing more than arbitrary curve-fits to the compressibility data; they simply characterize cracks by their "closure pressure". Models which are based on highly irregular grain-contact profiles could also be developed in analogy to the treatment given by Walsh and Grosenbaugh [1979] for rock fractures. Nevertheless, the assumption that sandstones contain a distribution of elliptical or spheroidal cracks has been shown to successfully explain many aspects of the mechanical behavior of sandstones, such as the change in Poisson's ratio with stress [Zimmerman and King, 1985]. Furthermore, these models can be used to relate mechanical properties to transport properties such as permeability [Seeburger and Nur, 1984] or thermal conductivity [Walsh and Decker, 1966; Zimmerman, 1989].

Many different approaches have been taken to the problem of relating the compressibility of a rock to the distribution of the aspect ratios of its pores. Morlier [1971] developed a direct relationship between the aspect ratio distribution function and the second derivative of the hydrostatic stress-strain curve [see also Walsh and Decker, 1966]. This method, however, uses the non-interactive dilute-crack approximation for the effective compressibility, so its applicability is restricted to rocks whose crack densities are less than about 0.20. Since Morlier's method can be thought of as yielding a first approximation to the actual aspect ratio distribution, and is simple to understand, it is worth presenting.

For rocks with low concentrations of penny-shaped cracks, eqn. (11.28) shows that $C_{bc} = C_r + 16(1-\nu_r^2)C_r\Gamma/9(1-2\nu_r)$. Note that any non-closable pores will contribute an additional term to C_{bc}; however, this compressibility will be independent of pressure, and will turn out to be of no consequence in the following analysis. At any given pressure, Γ must refer to the density of *open* cracks. If the cracks are treated as penny-shaped spheroids, the cracks that are still open at an effective pressure P will be those cracks whose initial aspect ratios were greater than $2(1-\nu_r)P/\pi G = 4(1-\nu_r^2)C_r P/3\pi(1-2\nu_r)$. (Recall from eqn. (5.7) that the compressibility of a penny-shaped crack essentially depends only on the differential pressure $P = P_c - P_p$.) Differentiation of eqn. (11.28) with respect to P, and use of the chain rule, gives

$$\frac{dC_{bc}}{dP} = \frac{16(1-\nu_r^2)C_r}{9(1-2\nu_r)}\frac{d\Gamma}{dP}$$

$$= \frac{16(1-\nu_r^2)C_r}{9(1-2\nu_r)}\frac{d\Gamma}{d\alpha}\frac{d\alpha}{dP}$$

$$= \frac{4\pi}{3}\left[\frac{4(1-\nu_r^2)C_r}{3\pi(1-2\nu_r)}\right]^2\frac{d\Gamma}{d\alpha}, \tag{12.1}$$

since $d\alpha/dP = 4(1-\nu_r^2)C_r/3\pi(1-2\nu_r)$. Eqn. (12.1) can be solved for $d\Gamma/d\alpha$, which is the negative of the crack aspect ratio distribution function $\gamma(\alpha)$:

$$\gamma(\alpha) = \frac{-d\Gamma}{d\alpha} = \frac{-3}{4\pi}\left[\frac{3\pi(1-2\nu_r)}{4(1-\nu_r^2)C_r}\right]^2\left[\frac{dC_{bc}}{dP}\right]_{P=3\pi(1-2\nu_r)\alpha/4(1-\nu_r^2)C_r}. \tag{12.2}$$

Since $\Gamma(\alpha)$ represents the number of cracks whose aspect ratios are *greater* than α, it is not quite the same as the cumulative distribution function; it is therefore convenient to work with the negative of the derivative of Γ. The notation of eqn. (12.2) indicates that the derivative dC_{bc}/dP must be evaluated at a pressure equal to $3\pi(1-2\nu_r)\alpha/4(1-\nu_r^2)C_r$. Aside from the multiplicative constant, the aspect ratio distribution function is equal to the derivative of the compressibility, which is to say the second derivative of the stress-strain curve itself.

As an example of the use of Morlier's expression (12.2), consider the exponentially decreasing compressibility functions that have been found [Wyble, 1958; Zimmerman, 1984a] to represent the behavior of many sandstones:

$$C_{bc} = C_{bc}^{\infty} + (C_{bc}^{i} - C_{bc}^{\infty})e^{-P/\hat{P}} \,. \tag{12.3}$$

C_{bc}^{i} is the compressibility at zero pressure, while C_{bc}^{∞} is the asymptotic compressibility that is reached at high pressures when all of the cracks are closed. $\hat{P}$ is a scaling factor, with dimensions of pressure, that determines the rate at which the compressibility levels off. Substitution of eqn. (12.3) into eqn. (12.2) yields

$$\gamma(\alpha) = \left[\frac{9(1-2\nu_r)}{16(1-\nu_r^2)C_r}\right]\frac{C_{bc}^{i} - C_{bc}^{\infty}}{\hat{\alpha}}\,e^{-\alpha/\hat{\alpha}} \,, \tag{12.4}$$

where, as in eqn. (12.2), P has been replaced with $3\pi(1-2\nu_r)\alpha/4(1-\nu_r^2)C_r$.

It is convenient to introduce another aspect ratio distribution function, $c(\alpha)$, which can be defined as follows. Recall that the total "density" of cracks whose initial aspect ratios lie between α and $\alpha + d\alpha$ is equal to $\gamma(\alpha)d\alpha$. Since $\Gamma = Nr^3/V_b$, and the volume of a spheroid of aspect ratio α is $4\pi r^3\alpha/3$, the porosity represented by these cracks is $4\pi\alpha\gamma(\alpha)d\alpha/3$. Hence the aspect ratio distribution function of the porosity can be written as

$$c(\alpha) = \frac{4\pi\alpha}{3}\gamma(\alpha) = \frac{3\pi(1-2\nu_r)}{4(1-\nu_r^2)C_r}[C_{bc}^{i} - C_{bc}^{\infty}]\,\frac{\alpha}{\hat{\alpha}}\,e^{-\alpha/\hat{\alpha}} \,. \tag{12.5}$$

Since $d(\alpha e^{-\alpha/\hat{\alpha}})/d\alpha = [1-\alpha/\hat{\alpha}]e^{-\alpha/\hat{\alpha}}$, the function $c(\alpha)$ has a maximum at $\alpha = \hat{\alpha}$. Since $\hat{P}$ is usually on the order of 1000 psi, and C_r for most minerals is on the order of 1×10^{-7}/psi, $\hat{\alpha}$ will be on the order of 1×10^{-4}. This is somewhat smaller than the "average" aspect ratio of most crustal rocks, as estimated by electron microscopy [Hadley, 1976] or other means. The discrepancy can be accounted for by realizing that since Morlier's method ignores "interactions" between cracks, it underestimates the pore compressibility (see Chapter 11); this in turns implies an underestimation of the initial aspect ratios of the cracks.

Exponential compressibility functions were used by Zimmerman [1984a] to fit measured C_{pc} pore compressibilities for three consolidated sandstones, Boise, Berea, and Bandera. The Boise is a medium-fine grained feldspar-quartzose sandstone with angular to sub-rounded grains, the Berea is a fine-grained quartzose sandstone with sub-angular to sub-rounded grains, and the Bandera is a very-fine grained quartzose sandstone with sub-angular to sub-rounded grains. The average grain diameters for these three sandstones have been estimated by Greenwald [1980] to be 300 μm for the Boise, 200 μm for the Berea, and 80 μm for the Bandera. If eqn. (2.7) is used to convert pore compressibilities to bulk compressibilities, the compressibility equations for these three sandstones are (see Fig. 12.1)

Boise:

$$C_{bc}(P) = 0.95\times10^{-4} + 2.79\times10^{-4}e^{-P/7.01\,\mathrm{MPa}}\ /\mathrm{MPa}\ ,$$

$$C_r = 0.251\times10^{-4}/\mathrm{MPa}\ ,$$

$$\nu_r = 0.188\ ; \tag{12.6}$$

Berea:

$$C_{bc}(P) = 1.05\times10^{-4} + 6.35\times10^{-4}e^{-P/4.74\,\mathrm{MPa}}\ /\mathrm{MPa}\ ,$$

$$C_r = 0.222\times10^{-4}/\mathrm{MPa}\ ,$$

$$\nu_r = 0.218\ ; \tag{12.7}$$

Bandera:

$$C_{bc}(P) = 0.82\times10^{-4} + 5.35\times10^{-4}e^{-P/8.33\,\mathrm{MPa}}\ /\mathrm{MPa}\ ,$$

$$C_r = 0.226\times10^{-4}/\mathrm{MPa}\ ,$$

$$\nu_r = 0.210\ . \tag{12.8}$$

Using the correspondence $P \rightarrow 3\pi(1-2\nu_r)\alpha/4(1-\nu_r^2)C_r$, eqns. (12.6-12.8) can be combined with eqns. (12.2) and (12.5) to yield the following aspect ratio distribution functions:

$$\text{Boise:}\quad c(\alpha) = 1.47\times10^{5}\,\alpha\exp(-\alpha/1.15\times10^{-4})\ , \tag{12.9}$$

$$\text{Berea:}\quad c(\alpha) = 5.29\times10^{5}\,\alpha\exp(-\alpha/7.54\times10^{-5})\ , \tag{12.10}$$

$$\text{Bandera:}\quad c(\alpha) = 2.57\times10^{5}\,\alpha\exp(-\alpha/1.31\times10^{-4})\ . \tag{12.11}$$

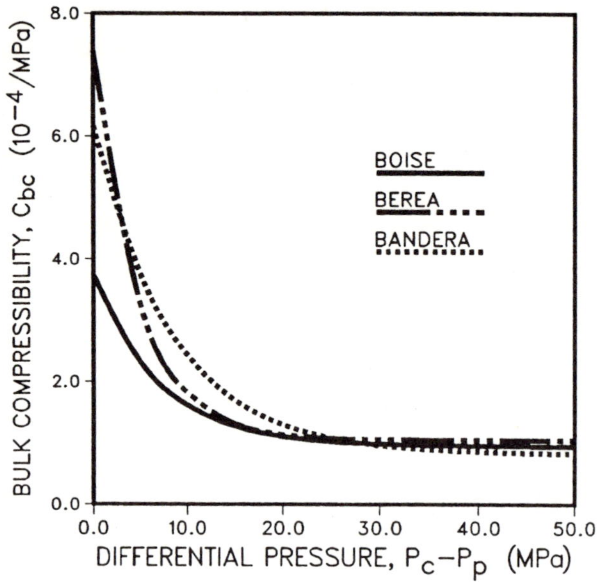

Fig. 12.1. Bulk compressibilities C_{bc} of three consolidated sandstones, as functions of the differential pressure, according to eqns. (12.6-12.8) [after Zimmerman, 1984a].

The aspect ratio distributions predicted by Morlier's method are plotted in Fig. 12.2. Note that the total crack porosity at zero pressure is equal to the integral of $c(\alpha)$ from $\alpha=0$ out to $\alpha=1$. Since the exponential terms will be negligible for $\alpha>1$, the integrals can mathematically be taken from 0 to ∞. The total crack porosity ϕ_{crack} can then be expressed in terms of the numerical coefficients in the distribution function. If $c(\alpha) = A\,\alpha\exp(-\alpha/\hat{\alpha})$, then

$$\phi_{crack} = \int_0^\infty c(\alpha)d\alpha = \int_0^\infty A\,\alpha\, e^{-\alpha/\hat{\alpha}}\,d\alpha = A\,\hat{\alpha}^2 . \qquad (12.12)$$

The total crack porosity for the three sandstones is therefore found to be 0.19% for the Boise, 0.30% for the Berea, and 0.44% for the Bandera. These values are in the same range as those which were estimated by Brower and Morrow [1985] for some low permeability gas sands, which was about 0.5%. However, more detailed analysis (see below) shows that Morlier's method usually overestimates the amount of crack porosity. The problem with Morlier's method stems from the fact that it ignores the pore-pore interaction effects that tend to cause the compressibility of a pore to exceed $4(1-\nu_r^2)C_r/3\pi(1-2\nu_r)\alpha$. Since the compressibility of

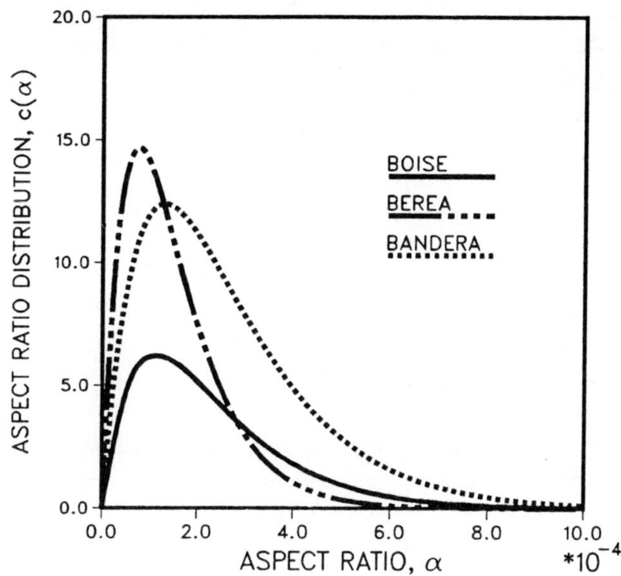

Fig. 12.2. Aspect ratio distributions of the three sandstones whose compressibilities are shown in Fig. 12.1. The distributions, which are given in eqns. (12.9-12.11), are calculated by Morlier's [1971] method.

each crack is underestimated, the aspect ratios are underestimated. However, the use of the no-interaction method for the effective moduli leads to an overestimation of the crack density Γ (see Fig. 11.6). The combination of these two errors typically causes the peak of the $c(\alpha)$ function to be shifted towards smaller values of α.

If interactions are accounted for, the compressibility of any individual crack will depend on the number of *open* nearby cracks. Since the open crack density depends on pressure, the compressibility, and hence the closing pressure, of each pore will vary with pressure. This effect greatly complicates the analysis needed to extract aspect ratio distributions from stress-strain data. Cheng and Toksöz [1979] developed an algorithm for finding aspect ratio distributions that attempts to account for pore-pore interactions. Their method assumes that the two acoustic wave velocities, V_T and V_L, have been measured on a dry specimen at a finite number of differential pressures. These velocities in turn yield the elastic moduli through the equations $G = \rho V_T^2$ and $K = \rho(V_L^2 - 4V_T^2/3)$. It is then assumed that at zero pressure, the rock contains pores whose aspect ratios are distributed among a finite number of values, say α_i, where i runs from 1 to N. At zero pressure, the concentrations of pores at each of these aspect ratios, $c(\alpha_i)$, is initially unknown, but must lead to the correct elastic moduli, according to some effective moduli theory. If the pressure is increased to some nonzero value, the aspect ratio populations change (according to the pore compressibility equations), but they still must lead to the correct values for the elastic moduli at this new pressure. By constraining the elastic moduli to fit the measured values, equations can be developed for the unknown values $c(\alpha_i)$. This leads to a system of nonlinear equations with N unknowns, which can be solved by a Newton-Raphson method. Cheng and Toksöz took N to be on the order of 10, and determined aspect ratio distributions for Navajo, Berea, and Boise sandstones, as well as a limestone, granite and marble.

Zimmerman [1984a] developed a methodology for extracting aspect ratio distributions from stress-strain curves that avoids the extensive numerical computations required by the Cheng-Toksöz method, and also has the advantage that it is not restricted to a finite number of aspect ratios. Since the derivation of this method is somewhat complicated, it is worthwhile to outline the basic steps involved before presenting them in detail. First, assume that experimental data exists that relates one of the porous rock compressibilities to the effective confining pressure P. Since all of the porous rock compressibilities are related to each other through the equations developed in Chapter 2, it can be assumed that C_{bc} is known experimentally. An effective moduli theory of the type discussed in Chapter 11 can then be used to relate C_{bc} to the density of *open* cracks, Γ. If C_{bc} is then eliminated between the $C_{bc}(P)$ and $C_{bc}(\Gamma)$ relations, the result is a relation between P and Γ. The aspect ratio α of each crack can then be related to the pressure required to close that crack by integrating the pore compressibility as a function of pressure. This yields a relationship between P and α, where P is the minimum pressure needed to close a crack of initial aspect ratio α. Finally, the pressure can be eliminated between the $P-\Gamma$ and $P-\alpha$ relations, yielding the change in open crack density as a function of aspect ratio, which is the desired result.

Although the implementation of this method requires choice of an effective moduli theory, the integral equation that expresses the relationship between the stress-strain curve and the aspect ratio distribution can be derived in a general form. Consider first the expression which defines the pore strain of a crack of aspect ratio α:

$$d\varepsilon_p = \frac{dV_p(\alpha)}{V_p^i(\alpha)} = -C_{pc}(P',\alpha)dP' . \tag{12.13}$$

Since the pressure in eqn. (12.13) is not intended to refer to the closing pressure for a crack of aspect ratio α, a dummy variable P' is used to avoid confusion. $C_{pc}(P',\alpha)$ represents the pore compressibility of a fixed pore of (initial) aspect ratio α, and is *not* the pore compressibility of the material as a whole. This notation is designed to indicate that this compressibility is in general a function of both the aspect ratio α and the pressure P'. The implicit dependence on pressure is due to the fact that the pressure determines the density of *open* cracks, which in turn affects the effective compressibility. Eqn. (12.13) can be integrated from $P'=0$ to $P'=P(\alpha)$, where $P(\alpha)$ is the minimum pressure required to close a crack of aspect ratio α. The limits of integration for ε_p are 0 and -1, since the pore strain is zero when the pressure is zero, and equals -1 at the point of closure. Therefore

$$\int_0^{-1} d\varepsilon_p = -\int_0^{P(\alpha)} C_{pc}(P',\alpha)dP',$$

so

$$1 = \int_0^{P(\alpha)} C_{pc}(P',\alpha)dP'. \tag{12.14}$$

Eqn. (12.14) implicitly defines a relationship between P and α, where P is identified as the pressure needed to close up a crack of initial aspect ratio α. Note that while the closing pressure is known *a priori* if the dilute-crack approximation (11.28) is used, in general the closing pressure $P(\alpha)$ must be found as part of the process of finding the aspect ratio distribution, since C_{pc} varies with pressure.

The pore compressibility C_{pc} of the entire body can be expressed in terms of the pore compressibilities of the individual pores. Consider first a finite number of pores, each with its own (possibly different) aspect ratio, and let $V_p^i(\alpha)$ represent that portion of the initial pore volume which is associated with pores of aspect ratio α. The total initial pore volume is simply the sum of $V_p^i(\alpha)$ over all α, *i.e.* $V_p^i=\sum V_p^i(\alpha)$, and pore volume increments can be expressed as $dV_p=\sum dV_p(\alpha)$. But $dV_p(\alpha)=-C_{pc}(P',\alpha)V_p^i(\alpha)dP'$ by definition, so that $dV_p=-\sum C_{pc}(P',\alpha)V_p^i(\alpha)dP'$. In terms of the overall pore compressibility, though, $dV_p=-C_{pc}V_p^i dP'$, so it is seen that $C_{pc}V_p^i=\sum C_{pc}(P',\alpha)V_p^i(\alpha)$. Dividing through by the initial bulk volume V_b^i yields the relation $C_{pc}\phi^i=\sum C_{pc}(P',\alpha)c(\alpha)$, where $c(\alpha)$ is the initial porosity associated with pores of aspect ratio α. For the more general case where the rock may have pores of any aspect ratio, the summation must be replaced by an integration, leading to

$$\phi^i C_{pc}(P) = \int_\alpha^1 c(\alpha')C_{pc}(P,\alpha')d\alpha', \tag{12.15}$$

where $c(\alpha)$ must now be considered to be a distribution function, and α' is a dummy variable of integration. The domain of integration includes the aspect ratios of all cracks which will still be open when the pressure equals P. If the distribution function $c(\alpha)$ is replaced by $4\pi\alpha\gamma(\alpha)/3$, we have

$$\phi^i C_{pc}(P) = \int_{\alpha}^{1} \frac{4\pi}{3}[\alpha' C_{pc}(P,\alpha')]\gamma(\alpha')d\alpha' \ . \tag{12.16}$$

The crucial step is to realize that the product $\alpha' C_{pc}(P,\alpha')$ should be *independent* of α', even though it varies with P. First, note that when the concentration of cracks is very dilute, this follows from eqn. (10.5):

$$\alpha' C_{pc}(P,\alpha') = \frac{4(1-\nu_r^2)C_r}{3\pi(1-2\nu_r)} \neq \text{a function of } \alpha' \ . \tag{12.17}$$

To see that this must be true at non-dilute crack concentrations, at least as a statistical average over large numbers of cracks, note that if the cracks are assumed to be randomly located with no correlations between their aspect ratios, they will all behave as if they were embedded in the same ''effective'' medium. Hence each crack will have a compressibility of the form given by eqn. (10.5), with C_r and ν_r replaced by some effective values. The appropriate effective moduli will change with pressure, but the important point is that at any given pressure, $C_{pc}(P,\alpha')$ will be inversely proportional to α'. (Since the energy associated with a pressurized crack is proportional to the product of its initial volume and its compressibility (see eqn. (11.26)), and the initial volume is proportional to α, this is equivalent to the statement that the crack energy is independent of α.) Thus, the combination $\alpha' C_{pc}(P,\alpha')$ can be brought outside the integral in eqn. (12.16):

$$\phi^i C_{pc}(P) = \frac{4\pi}{3}[\alpha C_{pc}(P,\alpha)]\int_{\alpha}^{1} \gamma(\alpha')d\alpha' \ . \tag{12.18}$$

Since the bracketed product in eqn. (12.18) is independent of the value of α', it can be replaced by $\alpha C_{pc}(P,\alpha)$. The integral appearing in eq. (12.18) is merely $\Gamma(\alpha)$, the cumulative density of all cracks with aspect ratios greater than α, which is by definition the total density of the cracks which are still open at pressure P. Therefore,

$$\phi^i C_{pc}(P) = \frac{4\pi}{3}[\alpha C_{pc}(P,\alpha)]\Gamma(\alpha) \ . \tag{12.19}$$

If eqn. (12.14) is now multiplied through by α, the result is

$$\alpha = \int_{0}^{P(\alpha)} [\alpha C_{pc}(P',\alpha)]dP' \ . \tag{12.20}$$

The product $[\alpha C_{pc}(P',\alpha)]$ can be eliminated between eqns. (12.19) and (12.20), yielding

$$\alpha = \frac{3\phi^i}{4\pi}\int_0^{P(\alpha)} \frac{C_{pc}(P')dP'}{\Gamma} . \tag{12.21}$$

Regardless of which effective moduli theory is chosen, C_{bc} will be a monotonically decreasing function of Γ. Since C_{bc} will also be a decreasing function of P, Γ can be used instead of P as the variable of integration in eq. (12.21). Formally, this is accomplished by replacing dP' with $[dP'(\Gamma)/d\Gamma]d\Gamma$, and replacing the limits of integration with $\Gamma(P'=0)$ and $\Gamma(P'=P)$. Although P will not be known explicitly as a function of Γ, the chain rule can be used to express $dP'/d\Gamma$ as $(dP'/dC_{bc})(dC_{bc}/d\Gamma)$. Since P is identified as the minimum pressure needed to close a crack of aspect ratio α, all cracks with higher aspect ratios higher than α will be open at pressure P, while all those with smaller aspect ratios are closed. Hence when $P=0$, $\alpha=0$. Therefore the limits of integration can also be thought of as $\Gamma(\alpha=0)$ and $\Gamma(\alpha)$:

$$\alpha = \frac{3\phi^i}{4\pi}\int_{\Gamma(0)}^{\Gamma(\alpha)} \frac{C_{pc}(\Gamma)\dfrac{dC_{bc}}{d\Gamma}}{\dfrac{dC_{bc}}{dP}\Gamma}\, d\Gamma . \tag{12.22}$$

Since $\phi^i C_{pc} = C_{bc} - C_r$ (see eqn. (2.7)), this can be rewritten as

$$\alpha = \frac{3}{4\pi}\int_{\Gamma(0)}^{\Gamma(\alpha)} \frac{[C_{bc} - C_r]\dfrac{dC_{bc}}{d\Gamma}}{\dfrac{dC_{bc}}{dP}\Gamma}\, d\Gamma . \tag{12.23}$$

$\Gamma(0)$ is the density of open cracks at zero pressure, hence it represents the density of non-closable pores. The precise distribution of aspect ratios of the non-closable pores cannot be found by Zimmerman's method. (The method of Cheng and Toksöz [1979] is capable of determining a *non-unique* aspect ratio distribution for the non-closable pores.) The value of $\Gamma(0)$ is found by inverting the $C_{bc}(\Gamma)$ relation, using the value $C_{bc} = C_{bc}(P=0)$. Eqn. (12.23) gives α as a function of Γ, from which can be found $c(\alpha) = 4\pi\alpha\gamma(\alpha)/3 = -4\pi\alpha\Gamma'(\alpha)/3$.

As an example of the use of eqn. (12.23), and the method of Cheng and Toksöz [1979], consider the measurements by Johnston [1978] on the elastic wave velocities of a dry Navajo sandstone (Fig. 12.3). Cheng and Toksöz [1979] used their algorithm to compute the aspect ratio distribution function of the porosity, which in their method emerges as a series of "delta"

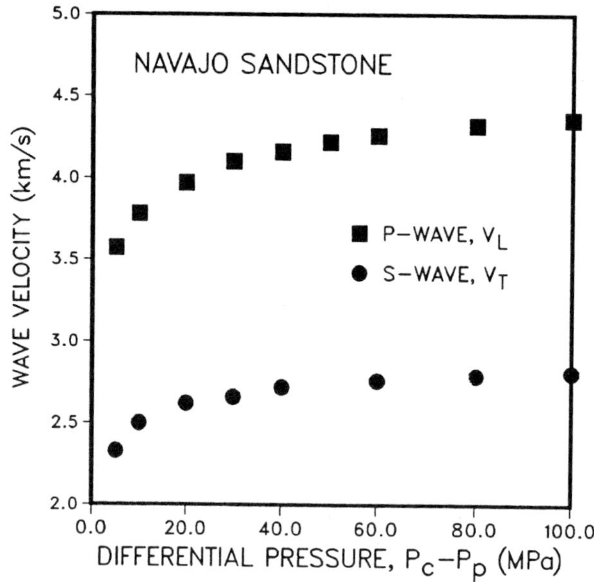

Fig. 12.3. Velocities of compressional and shear waves in a dry Navajo sandstone, as a function of differential pressure [after Johnston, 1978].

functions that are located at a predetermined set of aspect ratios, α_i. Since there is no unambiguous way of plotting a set of delta functions, it is convenient to use the integral of $c(\alpha)$, which will be denoted by $C(\alpha)$. This cumulative porosity distribution function $C(\alpha)$ represents the total amount of the initial porosity that is contained in pores whose aspect ratios are *less* than α. The results of their analysis are shown in Fig. 12.4.

In order to use eqn. (12.23), the bulk compressibility C_{bc} is first computed from the relationship $1/C_{bc} = K = \rho[V_L^2-(4/3)V_T^2]$, and is then fit to a curve of the form (12.3). Of course, any form of the $C_{bc}(P)$ relation could be used, including merely a table of values. If the predictions of an effective moduli theory, such as eqns. (11.36) and (11.37), are used, $dC_{bc}/d\Gamma$ can then be calculated, and eqn. (12.23) can be integrated. The results are also shown in Fig. 12.4. Note that the two methods yield results that are quite similar, although the method of Cheng and Toksöz predicts a slightly larger crack porosity. This apparent overestimation of the porosity can be ascribed to the use of pore and bulk compressibilities that are not consistent with eqn. (2.7).

Other approaches have been taken to the problem of determining aspect ratio distributions. Saito and Abe [1984] used the Kuster-Toksöz [1974] equations to study the differences

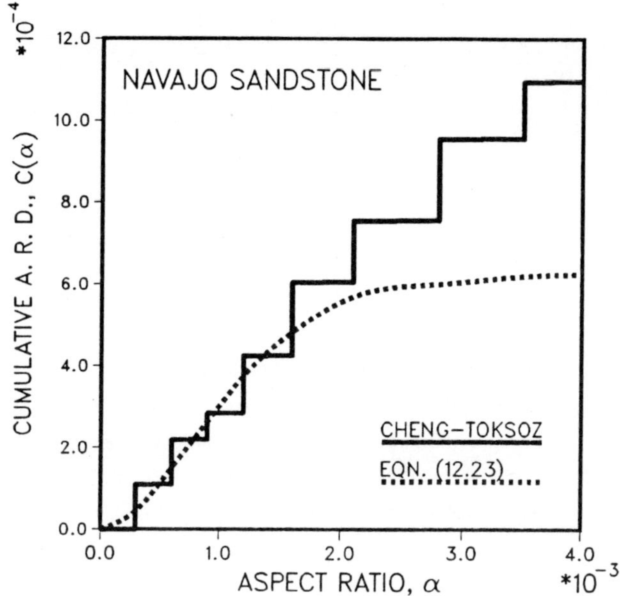

Fig. 12.4. Cumulative aspect ratio distribution functions for the "closable" porosity in a Navajo sandstone (see Fig. 12.3), according to the method of Cheng and Toksöz [1979] and eqn. (12.23).

between the elastic wave velocities in dry and water-saturated specimens. The aspect ratios were assumed to be distributed among eleven different aspect ratios, equally spaced along a logarithmic scale from 10^0 to 10^{-5}. Calculations were made for 3640 different aspect ratio distributions, and experimental results were then compared to these results, with the best match determining the appropriate distribution. Seeburger and Nur [1984] studied the effect of aspect ratio distribution on the mechanical and hydraulic properties of Berea, St. Peter, and Massilon sandstones. They assumed that all pores were two-dimensional tubes of elliptical cross-section, and used aspect ratios distributions derived by the Cheng-Toksöz method to predict the variation of permeability with pressure. Thomsen [1985] presented a scheme based on the assumption that all of the voids in a rock are either spherical pores or thin cracks; his scheme has the advantage of being explicitly consistent with the Gassmann relationship between dry and saturated compressibilities, eqn. (6.5).

PART THREE: COMPRESSIBILITY MEASUREMENTS

Chapter 13. Laboratory Measurements of Compressibilities

In order to measure the various porous rock compressibilities in a laboratory, it is necessary to be able to subject a sandstone specimen to controllable levels of confining and pore pressures, and to measure the resulting pore and bulk volume changes. A typical system for carrying out these measurements is the one used by Greenwald [1980], which was later modified by Zimmerman [1984a]. This system is illustrated schematically in Fig. 13.1. Roughly speaking, it consists of four subsystems: a pressure vessel, a cylindrical specimen and its various fittings, a confining pressure system (left side of Fig. 13.1), and a pore pressure system (right side of Fig. 13.1). This system will now be described in some detail, after which the actual measurement processes will be discussed.

The pressure vessel used by Greenwald [1980] and Zimmerman [1984a] was a hollow chrome-steel alloy cylinder with walls whose thickness was at least 2.5 cm. The top of the vessel screwed into the bottom piece, with an O-ring providing the pressure seal. The standard working pressure of this cell was 69 MPa (10,000 psi), although it would be advantageous to be able to pressurize the system to higher pressures. The internal length of the sealed vessel was about 23 cm. The bottom of the vessel had fittings to allow strain-gauge leads, thermocouple wires, etc., to enter the vessel. This simple vessel allowed hydrostatic confining pressures only to be applied to the cores; triaxial loading requires modifications (see below) such as that of Andersen [1988]. The cylindrical vessel is surrounded by two semi-cylindrical heaters, which are in turn surrounded by layers of insulation. The heaters are controlled by a temperature controller, which allows operation of the system at elevated temperatures. Insulation is used to provide a longer thermal time-constant for the system, which allows easier temperature control. An alternative is to have smaller heaters inside the vessel, surrounding the core, thus eliminating the need to heat up the entire system. Measurements of compressibilities under elevated temperatures have been made by Contreras *et al.* [1982] and Somerton [1982].

The vessel is filled with a fluid that provides the confining pressure. In order for this fluid not to corrode the vessel, or to create short circuits for the electrical leads in the vessel, an inert, nonconductive fluid such as silicone oil is usually used. The fluid is pressurized by an air-actuated positive-displacement pump, which is connected to an air-compressor (lower left of Fig. 13.1). The exit line from the cell passes through a pressure transducer, which measures the confining pressure P_c, and then a pressure regulator, after which it passes back into the oil

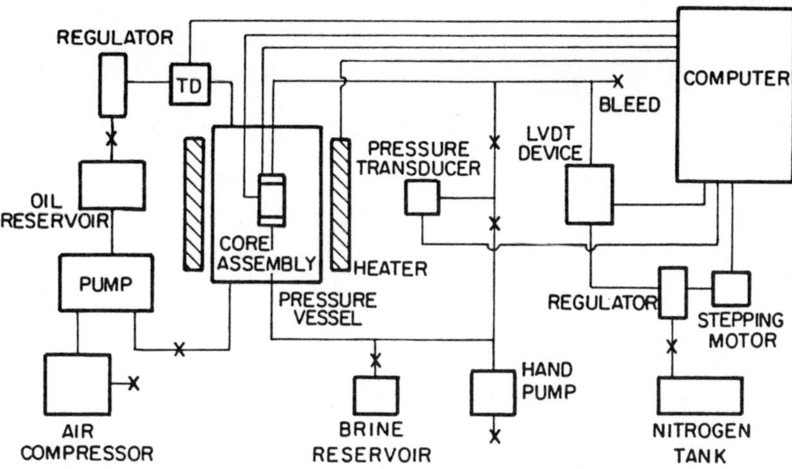

Fig. 13.1. Schematic diagram of system used to measure compressibilities of porous rocks [after Zimmerman, 1984a].

reservoir which supplies the pump. The reading from the pressure transducer is transmitted to a computer for data collection (upper right of Fig. 13.1); alternatively, the pressure could be read manually from a pressure gauge. The system used by Zimmerman [1984a] contained a strain-gauge type pressure transducer with a range of 0-20,000 psi (0-138 MPa), which measured the confining pressure to an accuracy of ± 5 psi.

The pore pressure system is indicated on the right side of Fig. 13.1. A strain-gauge type pressure transducer, with a range of 0-10,000 psi (0-69 MPa) and an accuracy of ± 5 psi, is used to measure the pore pressure. The device depicted in Fig. 13.2 is used both to measure changes in the pore volume, and to control the pore pressure. The device, consisting of a precision-bore cylinder containing a movable piston, is shunt-connected to the pore fluid system. Attached to this piston is a linear variable differential transformer (LVDT), whose motion allows volume changes to be measured to within $\pm 1.3 \times 10^{-4}\,\mathrm{cm}^3$. This piston is rigidly connected to a larger piston, with an area ratio of 20:1, upon which pressurized nitrogen acts to control the pore fluid pressure. The pressure of the nitrogen is controlled by a regulator that is operated by a stepping-motor, which in turn is controlled by the microcomputer.

The specimens used in most compressibility measurements are right circular cylinders, with diameters and length in the range of $1-2$ in ($2.5-5.0$ cm). It is of the utmost importance that the core be sheathed so as to keep the confining fluid and pore fluids from mixing. One such method for accomplishing this is the system used by Greenwald [1980] and Zimmerman *et al.* [1985a]. Cylindrical cores, 5.08 cm in length and 5.08 cm in diameter, were fitted with stainless-steel end-caps of the same diameter as the cores, which have holes drilled through them to allow for the passage of the pore fluid. A thin sheet of copper foil is then wrapped around the core, overhanging slightly onto the end-cap. Rubber O-rings are fitted into grooves

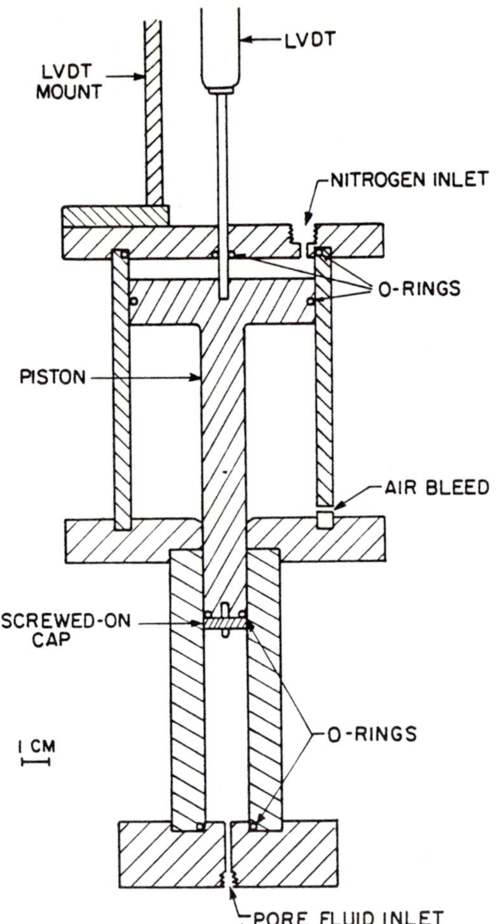

Fig. 13.2. Device used by Zimmerman [1984a] to measure changes in the pore volume, and to control the pore pressure. The area ratio of the two pistons is 20:1, so that the pore fluid pressure is twenty times the nitrogen pressure, which is controlled by the computer. The pore volume change is measured by the LVDT attached to the piston.

which are cut around the circumference of each end-cap, after which the entire core and end-cap assembly is sheathed in trifluorethylene heat-shrink tubing. A heat gun is used to shrink the tubing onto the core assembly. As a final precaution against leaks, a bead of silicone rubber is placed along the interface between the end-cap and the tubing [Greenwald, 1980].

The first part of the testing procedure is the fitting and saturation of the cores; this procedure is described in detail by Greenwald [1980]. After cutting the cores from a slab of sandstone, the faces are then milled square and flat. The specimen is then dried for twenty-four

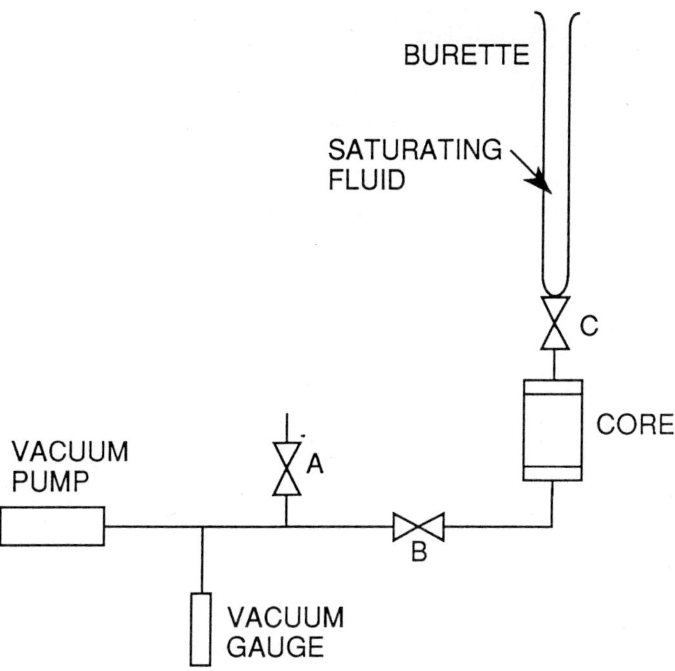

Fig. 13.3. Apparatus used to saturate the cores with the pore fluid [after Greenwald, 1980].

hours at 66°C in a vacuum oven to remove any moisture from the pores, after which it is carefully saturated with brine composed of 5 gms KCl per 1000 cm^3 of de-aerated, distilled water. Saturation is accomplished with an apparatus such as shown in Fig. 13.3. With valves A and C closed, a vacuum of about 100 μm Hg (13.3 Pa) is maintained on the core for about twelve hours, in order to avoid trapping any air in the pores. Valve B is then closed, and valve C is opened, allowing the brine to saturated the core. The saturation process is complete when the liquid level in the burette stabilizes. The pore volume of the core is then determined from the difference between the dry and saturated weight, while the bulk volume is determined by measuring the specimen's dimensions with vernier calipers.

Of the various porous rock compressibilities, by far the easiest to measure is C_{pc}, which is the derivative of the pore strain with respect to the confining pressure, with the pore pressure held constant. The earliest measurements of C_{pc} were made by Carpenter and Spencer [1940], using a method which is accurate and simple, but which is restricted to measurements in which the pore pressure is atmospheric. In their system, the pore pressure measuring device shown in Fig. 13.2 was in effect replaced by a graduated burette whose upper end is open to the atmosphere; a schematic version of this type of system is shown in Fig. 13.4. As the confining

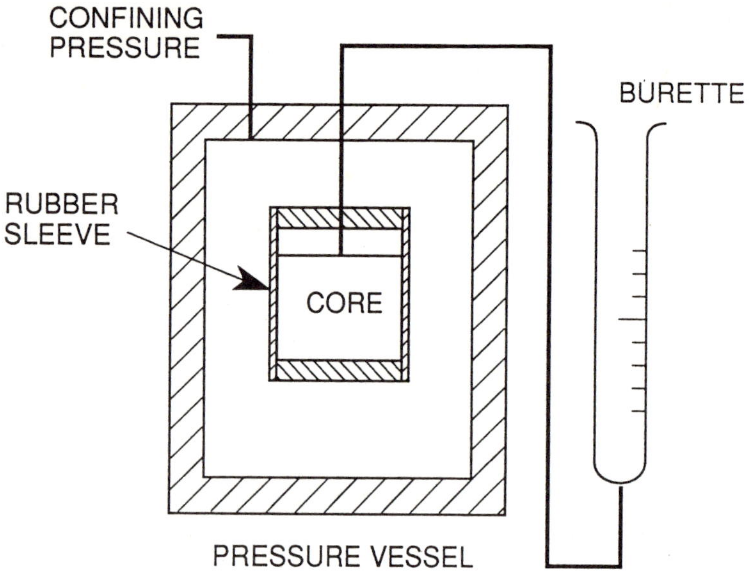

Fig. 13.4. Schematic diagram of system used to measure C_{pc} [after Lachance and Andersen, 1983]. As the confining pressure is increased, the pore fluid is expelled from the core into the burette.

pressure is increased, the pore volume of the core will decrease, and some of the pore fluid will be expelled into the burette. Since the pore fluid is at constant pressure (and temperature), the total volume of the pore fluid will remain unchanged. Hence any fluid expelled from the core will be reflected as an increase in the height of the fluid in the burette. Hughes and Cooke [1953], using a similar apparatus, placed a slug of mercury above the water column in the burette, to facilitate the location of the water level, and to prevent evaporation of the pore fluid.

Zimmerman [1984a] measured C_{pc} by a semi-automated method, with the device shown in Fig. 13.2 used in place of the burette. The measurement process commenced with the computer instructing the stepping motor to alter the setting of the regulator, so as to provide a particular level of nitrogen pressure. This in turn provides a particular pore fluid pressure by means of the previously discussed piston device. The confining pressure is then increased by injecting more silicone oil into the pressure vessel with the air-actuated pressure intensifier. Following a specified increment in the confining pressure, a period of several minutes is needed for the pore fluid system to equilibrate (which may be due to either frictional effects in the piston device or viscoelastic effects in the specimen), after which the computer records the

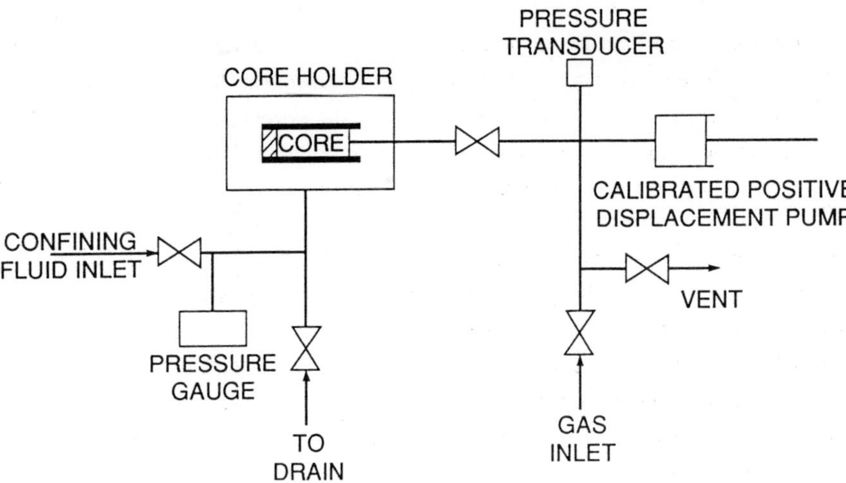

Fig. 13.5. Schematic diagram of the apparatus used by Sampath [1982] to measure C_{pc}, using nitrogen as the pore fluid.

pore volume change from the output of the LVDT. This is continued through a cycle of increasing and decreasing confining pressure; the procedure is then repeated at several different pore pressures. The confining pressure increments could also be automated, by utilizing a stepping motor and a screw-type piston pump.

In the method used by Sampath [1982], the pore fluid (nitrogen) is connected to a hand-operated positive displacement pump (see Fig. 13.5). After each increment in the confining pressure, the position of the piston in the pump is altered until the pore pressure returns to its (fixed) value; the pore volume change is read from the dial of the pump. Note that use of a gas as the pore fluid necessitates strict temperature control, else the thermal expansion of the gas overwhelm the volume change due to the pore compressibility. The advantage of this method is that by using nitrogen as a pore fluid, the difficulties associated with saturating the core with a liquid are avoided.

In order to measure C_{pp}, one must account for the fact that as the pore fluid pressure changes, the volume of the pore fluid will also change. For example, if the pore pressure is increased, the volume change measured by the pore pressure/pore volume device (Fig. 13.2) would partially reflect the fluid injected into the pore space of the core, but would also reflect the decrease of the pore fluid volume due to the fluid compressibility. Since the compressibility of the pore fluid is often of the same order as C_{pp}, this correction cannot be ignored. Another factor that must be taken into account is the ''compressibility'' of the measurement system itself. As the pore pressure increases, say, the volume available to the pore fluid outside of the vessel, in the tubes, pressure gauges, etc., will increase, due to the compliance of the apparatus. Fatt [1958] accounted for the compressibility of the pore fluid (kerosene, in his case) by estimating the amount of pore fluid in the entire system, and using published pressure-volume-temperature data. Greenwald [1980] attempted to account for both the compressibility of the pore fluid and the compliance of the system by using a published equation of state for the pore fluid (brine), and by running calibration tests without a core. The procedure needed to account for these extraneous effects is described in detail by Sawabini *et al.* [1971]. Another way to correct for these effects would be to place a hollow steel core in the vessel, and vary the pore pressure while holding the confining pressure constant, using a device such as that shown in Fig. 13.2. Since the steel core is less compressible than sandstone by a few orders of magnitude, it could be considered incompressible. The piston device would then be measuring the pressure-volume relationship of the experimental apparatus system itself. If the same test is then carried out with a real sandstone core, the discrepancy in the measurements of the pore fluid volume could be ascribed solely to changes in the pore volume of the core.

The two bulk compressibilities, C_{bc} and C_{bp}, can be measured in an apparatus of the type depicted in Fig. 13.1, if the cores are fitted with strain gauges. The process of fitting the cores with strain gauges is described by Greenwald [1980]; see Fig. 13.6. Since the surface of the core is in general too rough and irregular for strain gauges to be directly attached, high-temperature epoxy resin is first applied to a small region of the core. The epoxy is dried in an oven for twelve hours at 66 °C, after which it is sanded down to be flush with the core. The epoxy serves the purpose of filling in the surface pores, so as to provide a smooth surface for the application of the strain gauge. The epoxy also prevents the pore fluid from reaching the strain gauge, and causing a short circuit. In order for the strain gauge wires not to be short circuited or damaged by abrasion, a small patch of silicone rubber is applied at one end of the epoxy, for the gauges to be seated. The gauges are glued onto the epoxy with strain-gauge cement, and are then covered with a sheet of teflon. The cores are then further prepared in the manner described above. Typically, strain gauges are applied in both the longitudinal and transverse directions, so as to detect any anisotropy of the core. Since strain gauges often fail in the high pressure environment of the pressure vessel, it is prudent to use redundant gauges, in the event that one fails during a test. Greenwald [1980] also used a dummy gauge inside the vessel, not attached to a core, in order to compensate for the effects of pressure (or temperature).

Zimmerman [1984a] made measurements of C_{bp} on a Bandera sandstone core, using a method analogous to that used by Sampath [1982] to measure C_{pc}. This was done with a modified version of the experimental apparatus shown in Fig. 13.1, with a manually-controlled screw-type pump replacing the air-actuated pressure intensifier. This pump functions as the bulk volume analogue of the device depicted in Fig. 13.2, with the volumetric displacement read manually from a vernier-type scale instead of as the output of an LVDT. As the pore pressure is increased, say, the bulk volume of the core will increase. The confining fluid then

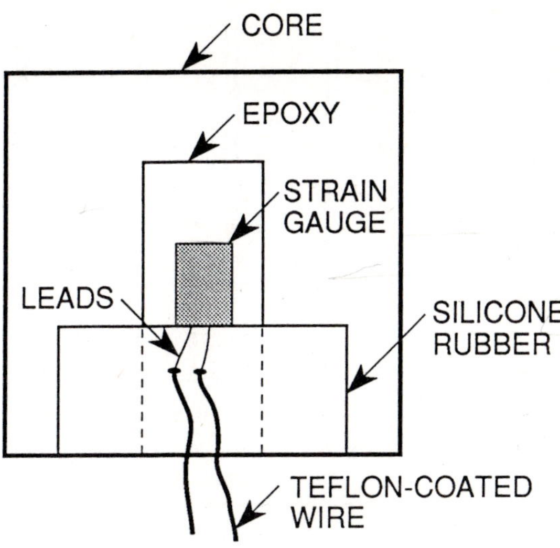

Fig. 13.6. Schematic diagram of method used by Greenwald [1980] to attach strain gauges to cores.

has less space to occupy, and its pressure increases slightly. The screw pump is then released, until the confining pressure returns to its nominal value. Since C_{bp} is measured at constant confining pressure, the volume of the confining fluid is constant, and so the bulk volume change of the core is measured directly by the vernier scale of the pump. These measurements proved to be very sensitive to temperature fluctuations of the system, due mainly to the large ratio of confining fluid volume to rock volume. This problem was partially mitigated by using a less compressible confining fluid (glycerin) for the C_{bp} measurements; an optimal system would have as small a ratio of confining fluid volume to rock sample volume as possible.

The measurements discussed above were all made under conditions of hydrostatic confining pressure. Triaxial stress measurements of sandstone compressibility were made by Chierici *et al.* [1967], Wilhelmi and Somerton [1967], Teeuw [1971], and Andersen [1988], using devices similar to that shown in Fig. 13.7. This apparatus allows a uniaxial stress to be applied by a piston, superimposed on top of the biaxial stress applied to the sides of the core by the ''annulus pressure'' (see Fig. 13.7). Andersen [1988] used strain gauges to measure the bulk strain along the longitudinal and transverse directions of the core, and used an external burette to measure the volume of pore fluid expelled from the core. If the core was linearly elastic, relating triaxial compressibilities to hydrostatic compressibilities would be trivial [see Teeuw, 1971]; for a typical nonlinear sandstone, this relationship is not yet entirely understood [see Andersen, 1988].

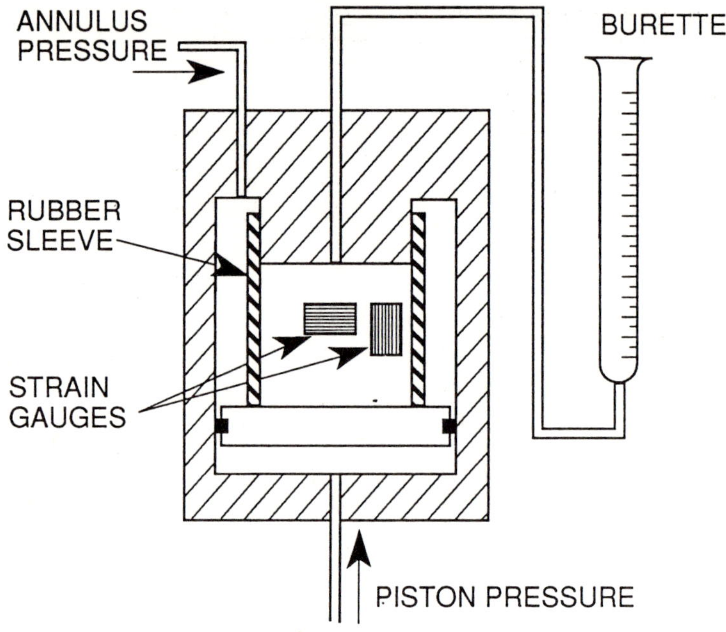

Fig. 13.7. Schematic diagram of apparatus used to measure pore and bulk compressibilities under triaxial loading conditions [after Andersen, 1988]. The confining fluid in the annulus provides two equal horizontal principal stresses, while the fluid acting on the piston provides the third principal stress. An O-ring is used to prevent pressure equilibration between these two fluids.

Undrained bulk compressibility can be measured by using the same procedure as used to measure C_{bc}, but with the pore fluid lines closed off. In order that the experimental set-up exactly reproduce "undrained" conditions, it is important that no pore fluid exist outside of the core. This is easily accomplished by plugging off the pore fluid lines at the end-caps. However, such an experimental configuration is not conducive to measuring the induced pore pressure, since these measurements require a pressure gauge of some sort to be in contact with the pore fluid. Excess volume in lines, gauges, etc., would provide a "sink" for the pore fluid, leading to an underestimation of the induced pore pressure coefficient B. Wissa [1969] analyzed the "undrained" compression of a core connected to external pore fluid tubing, and concluded that the ratio of pore fluid volume exterior to the core to the actual pore volume should not exceed 0.003. Green and Wang [1986] devised a system to measure B that placed a pore pressure transducer flush against the core (Fig. 13.8). The only extraneous volume was in the transducer chamber itself, and it typically amounted to only about 5×10^{-4} of the pore

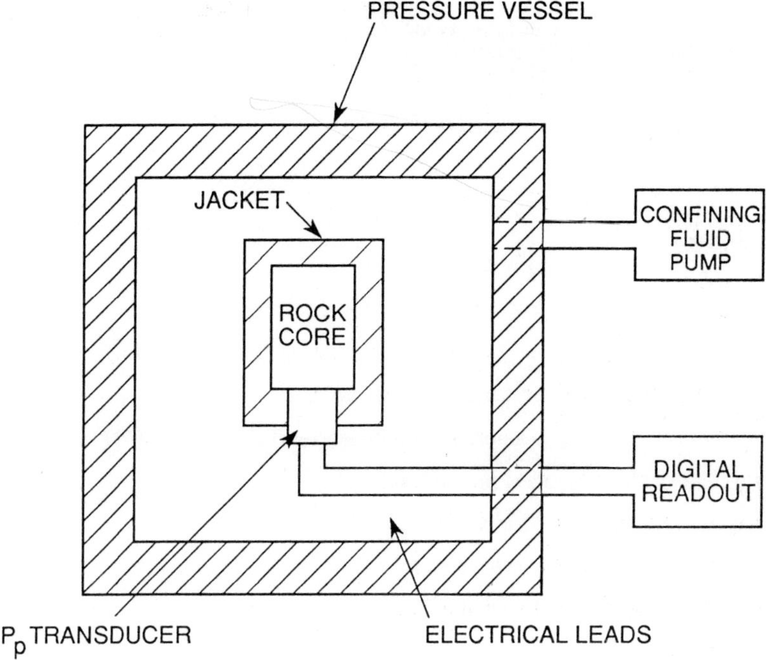

Fig. 13.8. Schematic diagram of apparatus used by Green and Wang [1986] to measure induced pore pressures during undrained compression.

volume of the core. This system was used to measured the induced pore pressure coefficient of Berea, Massillon, and Tunnel City sandstone cores. Measurements of induced pore pressures in porous rocks and sands have also been made by Mesri *et al.* [1976] and Dropek *et al.* [1978].

Nomenclature

Lower-case Arabic letters

a	Semi-major axis of elliptical pore [Chapter 8]
a_n	Coefficient in mapping function for two-dimensional pore [eqn. (8.7)]
b	Semi-minor axis of elliptical pore [Chapter 8]
b	Contact-region parameter for rectangular crack [eqn. (9.5)]
c	Closure parameter for cusp-shaped cracks [eqns. (9.9,9.15)]
c	Aspect ratio distribution function [Chapter 12]
e	Void ratio [eqn. (1.4)]
e	Eccentricity of elliptical pore [eqn. (8.6)]
f	Arbitrary function [Chapter 5]
g	Gravitational acceleration [Chapters 5,7]
h	Metric coefficient of elliptical coordinate system [eqn. (8.2)]
k	Permeability [Chapters 1,6,7]
m	Parameter in mapping function for hypotrochoid [eqn. (8.9)]
m_b	Effective stress coefficient for bulk compressibility [eqn. (4.1)]
m_p	Effective stress coefficient for pore compressibility [eqn. (4.2)]
$\mathbf{n}$	Outward unit normal vector [Chapter 7]
n	Parameter in mapping function for hypotrochoid [eqn. (8.9)]
n_b	Effective stress coefficient for bulk strain [eqn. (4.1)]
$\bar{n}_b$	Secant value of n_b [eqn. (5.18)]
n_p	Effective stress coefficient for pore strain [eqn. (4.2)]
$\bar{n}_p$	Secant value of n_p [eqn. (5.20)]
r	Semi-major axis of penny-shaped crack [Chapter 11]
s	Number of sides of hypotrochoidal pore [eqn. (8.11)]

t Time

$\mathbf{t}$ Traction vector [Chapter 7]

$\mathbf{u}$ Displacement vector [Chapter 7]

u_n Displacement normal to surface of elliptical pore [eqn. (8.1)]

u_i Displacement in ith-direction [Chapters 2,7]

w Crack face displacement (or half-width) [Chapter 9]

z Vertical coordinate [Chapters 5,7]

z Complex variable in the physical plane [Chapter 8]

Upper-case Arabic letters

A Curve-fit parameter for stress-strain relationship [eqn. (5.8)]

A Cross-sectional area of tubular pore [Chapter 8]

B Curve-fit parameter for stress-strain relationship [eqn. (5.8)]

B Induced pore pressure coefficient [eqn. (6.10)]

C Compressibility [eqn. (1.1)]

C Cumulative aspect ratio distribution function [Chapter 12]

$\overline{C}$ Effective compressibility of heterogeneous solid [Chapter 2]

C_{bc} Bulk compressibility [eqn. (1.5)]

$\overline{C}_{bc}$ Secant value of C_{bc} [eqn. (5.16)]

C_{bc}^{i} Bulk compressibility at zero pressure [eqn. (12.3)]

C_{bc}^{∞} Bulk compressibility at high pressures [eqn. (12.3)]

C_{bp} Bulk compressibility [eqn. (1.6)]

C_{pc} Pore compressibility [eqn. (1.7)]

$\overline{C}_{pc}$ Secant value of C_{pc} [eqn. (5.19)]

C_{pc}^{o} Pore compressibility of circular pore [eqn. (8.10)]

C_{pp} Pore compressibility [eqn. (1.8)]

C_f Compressibility of the pore fluid [Chapters 1,6,7]

C_i Compressibility of ith component [Chapter 2]

C_r Compressibility of the rock mineral phase

C_t Total compressibility of the pore space and fluid [Chapters 1,6,7]

C_u Undrained bulk compressibility [eqn. (6.5)]

C_{ϕ} Brown-Korringa compressibility [eqn. (6.7)]

D	Curve-fit parameter for stress-strain relationship [eqn. (5.8)]
E	Young's modulus
E	Complete elliptic integral of the second kind [Chapters 8,9]
$\mathbf{F}$	Force vector [Chapter 7]
G	Shear modulus
$\mathbf{I}$	Identity matrix [Chapter 7]
K	Bulk modulus
K	Complete elliptic integral of the first kind [Chapter 9]
L	Length of bar, element, or region of interest [Chapters 2,6,7]
M	Compressional wave modulus, $=K+4G/3$ [Chapter 11]
M	Generic elastic modulus [Chapter 11]
P	Pressure
P_c	Confining pressure
P_d	Differential pressure, $= P_c - P_p$
P_e	Effective pressure
P_p	Pore pressure
$\bar{P}$	Average pressure in the rock mineral phase [eqn. (4.13)]
P^*	Pressure at which an isolated elliptical crack closes [eqn. (9.3)]
$\hat{P}$	Pressure parameter for compressibility curve-fit [eqns. (5.8,12.3)]
P'	Dummy pressure variable for integration [Chapters 5,12]
R	Radius of curvature of pore [Fig. 8.5]
R	Spheroidal-coordinate system parameter [eqns. (10.1,10.4)]
S_w	Water saturation [Chapter 6]
$\mathbf{T}$	Stress matrix [Chapter 7]
T_{ij}	Component of stress matrix [Chapter 7]
T	Applied remote shear stress [Chapter 11]
V	Volume
V_b	Bulk volume
V_f	Pore fluid volume
V_L	Velocity of longitudinal (compressional) waves, $= \{[K+(4G/3)]/\rho\}^{1/2}$
V_p	Pore volume
V_r	Rock mineral volume [eqn. (1.2)]

V_T Velocity of transverse (shear) waves, $= (G/\rho)^{1/2}$

W Work done by applied loads [Chapter 2]

Greek letters

α Aspect ratio of elliptical or spheroidal pore [Chapters 5,8-12]

α Angular coordinate in ζ plane [Chapter 8]

$\hat{\alpha}$ Characteristic aspect ratio [Chapter 12]

β Compressibility [Chapter 1]

β Angular coordinate in elliptical coordinate system [eqns. (8.1,8.2)]

β_{pc} Normalized pore compressibility [eqn. (8.13)]

Δ Change in a variable

$\boldsymbol{\varepsilon}$ Strain matrix [Chapter 7]

ε_{ij} Component of strain matrix [eqn (7.6)]

ε Strain, $= \Delta V/V$

ε_b Bulk strain, $= \Delta V_b/V_b$

ε_p Pore strain, $= \Delta V_p/V_p$

ε_r Rock mineral phase strain, $= \Delta V_r/V_r$

γ Hole boundary in transformed ζ-plane [Chapter 8]

γ Crack aspect ratio distribution function [Chapter 12]

Γ Hole boundary in the physical z-plane [Chapter 8]

Γ Crack density parameter, $= Nr^3/V_b$ [Chapters 11,12]

κ Compressibility [Chapter 1]

λ Elastic moduli parameter [eqns. (11.16,11.17)]

μ Viscosity of pore fluid [Chapters 1,2,7]

ν Poisson('s) ratio, $= (3K - 2G)/(6K + 2G)$

ϕ Porosity [eqn. (1.3)]

Π Perimeter of two-dimensional pore [eqn. (8.14)]

ρ Density

ρ_f Density of the pore fluid [Chapters 5,7]

ρ_r Density of rock mineral phase [Chapter 5]

ρ_t Total density of overburden [eqn. (5.25)]

ρ_w Density of water [eqn. (5.27)]

Σ Elastic strain energy [Chapter 11]

τ Time constant for pore pressure equilibration [eqn. (6.8)]

θ Dummy variable of integration [eqn. (8.15)]

χ Volume fraction of mineral [Chapter 2]

ζ Complex variable in the transformed plane [Chapter 8]

Subscripts, superscripts, etc.

c Confining

d Differential

D Deviatoric part of matrix [eqn. (7.5)]

e Effective

f Fluid

ϕ Porosity

H Voigt-Reuss-Hill average [Chapter 2]

i Initial value at zero stress

I Isotropic part of matrix [eqn. (7.3)]

ij Components of matrix [Chapter 7]

n Normal component

o Value for circular two-dimensional pore [Chapter 8]

p Pore

r Rock mineral phase

R Reuss estimate of elastic moduli [Chapter 2]

u Undrained compression [Chapter 6]

V Voigt estimate of elastic moduli [Chapter 2]

w Water

References

Adams, L. H., and Williamson, E. D., 1923. The compressibility of minerals and rocks at high pressures. *Journal of the Franklin Institute,* 195: 475-529.

Andersen, M. A., 1988. Predicting reservoir-condition pore volume compressibility from hydrostatic-stress laboratory data. *SPE Reservoir Engineering,* 3(3): 1078-1082.

Andersen, M. A., and Jones, F. O., Jr., 1985. A comparison of hydrostatic-stress and uniaxial-strain pore-volume compressibilities using nonlinear elastic theory. In: *Research and Engineering Applications in Rock Masses: Proceedings of the 26th U.S. Symposium on Rock Mechanics.* Balkema, Rotterdam, pp. 403-410.

Anderson, O. L., Schreiber, E., Liebermann, R., and Soga, N., 1968. Some elastic constant data on minerals relevant to geophysics. *Reviews of Geophysics and Space Physics,* 6(4): 491-524.

Barenblatt, G. I., 1962. The mathematical theory of equilibrium cracks in brittle fracture. In: *Advances in Applied Mechanics, Vol. 7.* Academic Press, New York, pp. 55-129.

Bernabe, Y., Brace, W. F., and Evans, B., 1982. Permeability, porosity, and pore geometry of hot-pressed calcite. *Mechanics of Materials,* 1(3): 173-183.

Berry, D. S., 1960. An elastic treatment of ground movement due to mining - I. Isotropic ground. *Journal of the Mechanics and Physics of Solids,* 8(4): 280-292.

Berryman, J. G., 1988. Seismic wave attenuation in fluid-saturated porous media. *Pure and Applied Geophysics,* 128(1): 423-432.

Berryman, J. G., 1989. Estimating effective moduli of composites using quantitative image analysis. In: R. V. Kohn and G. W. Milton (Editors), *Random Media and Composites.* Society for Industrial and Applied Mathematics, Philadelphia, pp. 3-12.

Billington, E. W., and Tate, A., 1981. *The Physics of Deformation and Flow.* McGraw-Hill, New York, 626 pp.

Biot, M. A., 1941. General theory of three-dimensional consolidation. *Journal of Applied Physics,* 12(2): 155-164.

Biot, M. A., 1956a. Theory of propagation of elastic waves in a fluid-saturated porous solid. I. Low-frequency range. *Journal of the Acoustical Society of America,* 28(2): 168-178.

Biot, M. A., 1956b. Theory of propagation of elastic waves in a fluid-saturated porous solid. II. Higher frequency range. *Journal of the Acoustical Society of America,* 28(2): 179-191.

Biot, M. A., 1973. Nonlinear and semilinear rheology of porous solids. *Journal of Geophysical Research,* 78(23): 4924-4937.

Biot, M. A., 1974. Exact simplified non-linear stress and fracture analysis around cavities in rock. *International Journal of Rock Mechanics and Mining Sciences,* 11(7): 261-266.

Black, A. D., Dearing, H. L., and DiBona, B. G., 1985. Effects of pore pressure and mud filtration on drilling rates in a permeable sandstone. *SPE Journal of Petroleum Technology,* 37(10): 1671-1681.

Brace, W. F., 1965. Some new measurements of linear compressibility of rocks. *Journal of Geophysical Research,* 70(2): 391-398.

Brandt, H., 1955. A study of the speed of sound in porous granular media. *ASME Journal of Applied Mechanics,* 22(4): 479-486.

Brower, K. R., and Morrow, N. R., 1985. Fluid flow in cracks as related to low-permeability gas sands. *SPE Journal,* 25(2): 191-201.

Brown, R. J. S., and Korringa, J., 1975. On the dependence of the elastic properties of a porous rock on the compressibility of the pore fluid. *Geophysics,* 40(4): 608-616.

Bruner, W. M., 1976. Comment on ''Seismic velocities in dry and cracked solids'' by Richard J. O'Connell and Bernard Budiansky. *Journal of Geophysical Research,* 81(14): 2573-2576.

Budiansky, B., 1965. On the elastic moduli of some heterogeneous materials. *Journal of the Mechanics and Physics of Solids,* 13(4): 223-227.

Budiansky, B., and O'Connell, R. J., 1976. Elastic moduli of a cracked solid. *International Journal of Solids and Structures,* 12(2): 81-97.

Carpenter, C. B., and Spencer, G. B., 1940. Measurements of compressibility of consolidated oil-bearing sandstones. *U.S. Bureau of Mines Report 3540,* 20 pp.

Carroll, M. M., and Katsube, N., 1983. The role of Terzaghi effective stress in linearly elastic deformation. *ASME Journal of Energy Resources and Technology,* 105(4): 509-511.

Chen, H.-S., and Acrivos, A., 1978. The effective elastic moduli of composite materials containing spherical inclusions at non-dilute concentrations. *International Journal of Solids and Structures,* 14(5): 349-364.

Cheng, C. H., and Toksöz, M. N., 1979. Inversion of seismic velocities for the pore aspect ratio spectrum of a rock. *Journal of Geophysical Research,* 84(13): 7533-7543.

Chierici, G. L., Ciucci, G. M., Eva, F., and Long, G., 1967. Effect of the overburden pressure on some petrophysical parameters of reservoir rocks. In: *Proceedings of the Seventh World Petroleum Congress, Vol. 2.* Elsevier, London, pp. 309-338.

Chilingarian, G. V., and K. H. Wolf, (Editors), 1975. *Compaction of Course-Grained Sediments, Vol. 1*. Elsevier, Amsterdam, 546 pp.

Clark, N. J., 1969. *Elements of Petroleum Reservoirs*. Society of Petroleum Engineers, Dallas, 250 pp.

Clark, S. P., Jr., (Editor), 1966. *Handbook of Physical Constants*. (Geological Society of America Memoir 97) Geological Society of America, New York, 587 pp.

Cleary, M. P., 1978. Elastic and dynamic response regimes of fluid-impregnated solids with diverse microstructures. *International Journal of Solids and Structures,* 14(10): 795-819.

Contreras, E., Iglesias, E., and Bermejo, F., 1982. Effects of temperature and stress on the compressibilities, thermal expansivities, and porosities of Cerro Prieto and Berea sandstones to 9000 psi and 280C. In: *Proceedings of the Eighth Workshop on Geothermal Reservoir Engineering*. Stanford University, pp. 197-203.

Delameter, W. R., Herrmann, G., and Barnett, D. M., 1975. Weakening of an elastic solid by a rectangular array of cracks. *ASME Journal of Applied Mechanics,* 42(1): 74-79.

Detournay. E., and Cheng, A. H.-D., 1988. Poroelastic response of a borehole in a non-hydrostatic stress field. *International Journal of Rock Mechanics and Mining Sciences* 25(3): 171-182.

Dewey, J. M., 1947. The elastic constants of materials loaded with non-rigid fillers. *Journal of Applied Physics,* 18(6): 578-581.

Digby, P. J., 1981. The effective elastic moduli of porous granular rocks. *ASME Journal of Applied Mechanics,* 48(4): 803-808.

Dobrynin, V. M., 1962. Effect of overburden pressure on some properties of sandstones. *SPE Journal,* 2(4): 360-366.

Domenico, S. N., 1977. Elastic properties of unconsolidated porous sand reservoirs. *Geophysics,* 42(7): 1339-1368.

Dropek, R. K., Johnson, J. N., and Walsh, J. B., 1978. The influence of pore pressure on the mechanical properties of Kayenta sandstone. *Journal of Geophysical Research,* 83(B6): 2817-2824.

Dundurs, J., 1967. Effect of elastic constants on stress in a composite under plane deformation. *Journal of Composite Materials,* 1(3): 310-322.

Economides, M. J., and Nolte, K. G., 1989. *Reservoir Stimulation, 2nd Ed.* Schlumberger, Houston, 426 pp.

Edwards, R. H., 1951. Stress concentrations around spheroidal inclusions and cavities. *ASME Journal of Applied Mechanics,* 18(1): 19-27.

Eshelby, J. D., 1957. The determination of the elastic field of an ellipsoidal inclusion, and related problems. *Proceedings of the Royal Society of London,* A241: 376-396.

Fatt, I., 1958a. Compressibility of sandstones at low to moderate pressures. *Bulletin of the AAPG,* 42(8): 1924-1957.

Fatt, I., 1958b. Pore volume compressibilities of sandstone reservoir rocks. *Petroleum Transactions of the AIME,* 213: 362-364.

Fatt, I., 1959. The Biot-Willis elastic coefficients for a sandstone. *ASME Journal of Applied Mechanics,* 26(2): 296-297.

Fil'shtinskii, L. A., 1964. Stresses and displacements in an elastic sheet weakened by a doubly-periodic set of equal circular holes. *Applied Mathematics and Mechanics,* 28(3): 530-543.

Fletcher, A., 1940. A table of complete elliptic integrals. *Philosophical Magazine, 7th Series,* 30(203): 516-519.

Gassmann, F., 1951a. Uber die Elastizität Poröser Medien (On the elasticity of porous media). *Vierteljahrsschrift der Naturforschenden Gesellschaft in Zürich,* 96(1): 1-23.

Gassmann, F., 1951b. Elastic waves through a packing of spheres. *Geophysics,* 15(4): 673-685.

Geertsma, J., 1957. The effect of fluid pressure decline on volumetric changes of porous rocks. *Petroleum Transactions of the AIME,* 210: 331-340.

Geertsma, J., 1973. Land subsidence above compacting oil and gas reservoirs. *SPE Journal of Petroleum Technology,* 25(6): 734-744.

Gilluly, J., and Grant, U. S., 1949. Subsidence in the Long Beach harbor area, California. *Bulletin of the Geological Society of America,* 60(3): 461-530.

Green, D. H., and Wang, H. F., 1986. Fluid pressure response to undrained compression in saturated sedimentary rock. *Geophysics,* 51(4): 948-956.

Greenwald, R. F., 1980. *Volumetric Response of Porous Media to Pressure Variations.* Ph.D. dissertation, University of California at Berkeley, 174 pp.

Griffith, A. A., 1920. The phenomena of rupture and flow in solids. *Philosophical Transactions of the Royal Society of London,* A221: 163-198.

Gudehus, G., (Editor), 1977. *Finite Elements in Geomechanics.* John Wiley & Sons, New York, 573 pp.

Hadley, K., 1976. Comparison of calculated and observed crack densities and seismic velocities in Westerly granite. *Journal of Geophysical Research,* 81(20): 3484-3494.

Hall, H. N., 1953. Compressibility of reservoir rocks. *Petroleum Transactions of the AIME,* 198: 309-311.

Hashin, Z., 1983. Analysis of composite materials - a survey. *ASME Journal of Applied Mechanics,* 50(3): 481-505.

Hashin, Z., 1988. The differential scheme and its application to cracked materials. *Journal of the Mechanics and Physics of Solids,* 36(6): 719-734.

Hashin, Z., and Shtrikman, S., 1961. Note on a variational approach to the theory of composite elastic materials. *Journal of the Franklin Institute,* 271(4): 336-341.

Henyey, F. S., and Pomphrey, N., 1982. Self-consistent elastic moduli of a cracked solid. *Geophysical Research Letters,* 9(8): 903-906.

Hill, R., 1952. The elastic behaviour of a crystalline aggregate. *Proceedings of the Physical Society of London,* A65: 349-354.

Hill, R., 1965. A self-consistent mechanics of composite materials. *Journal of the Mechanics and Physics of Solids,* 13(4): 213-222.

Horii, H., and Nemat-Nasser, S., 1985. Elastic fields of interacting homogeneities. *International Journal of Solids and Structures,* 21(7):731-745.

Hughes, D. S., and Cooke, C. E., Jr., 1953. The effect of pressure on the reduction of pore volume of consolidated sandstones. *Geophysics,* 18(2): 298-309.

Inglis, C. E., 1913. Stresses in a plate due to the presence of cracks and sharp corners. *Transactions of the Institute of Naval Architects,* 55: 219-230.

Jaeger, J. C., and Cook, N. G. W., 1979. *Fundamentals of Rock Mechanics, 3rd Ed.* Chapman and Hall, London, 593 pp.

Johnson, J. P., Rhett, D. W., and Siemers, W. T., 1989. Rock mechanics of the Ekofisk reservoir in the evaluation of subsidence. *SPE Journal of Petroleum Technology,* 41(7): 717-722.

Johnston, D. H., 1978. *The Attenuation of Seismic Waves in Dry and Saturated Rocks.* Ph.D. dissertation, Massachussetts Institute of Technology.

Kantorovich, L. V., and Krylov, V. I., 1958. *Approximate Methods of Higher Analysis.* Noordhoff, Groningen, 681 pp.

King, M. S., 1969. Static and dynamic moduli of rocks under pressure. In: *Rock Mechanics - Theory and Practice: Proceedings of the Eleventh U.S. Symposium on Rock Mechanics.* Society of Mining Engineers, New York, pp. 329-351.

Knopoff, L., 1963. The theory of finite strain and compressibility of solids. *Journal of Geophysical Research,* 68(10): 2929-2932.

Knutson, C. F., and Bohor, B. F., 1963. Reservoir rock behavior under moderate confining pressure. In: *Rock Mechanics: Proceedings of the Fifth Symposium on Rock Mechanics.* Pergamon, New York, pp. 627-659.

Koiter, W. T., 1959. An infinite row of collinear cracks in an infinite elastic sheet. In: *Ingenieur-Archiv: Festshcrift Richard Grammel.* Springer-Verlag, Berlin, pp. 168-172.

Kolosov, G. V., 1909. *On an Application of the Theory of Functions of a Complex Variable to a Plane Problem in the Mathematical Theory of Elasticity.* Ph.D. dissertation, Dorpat

University (in Russian).

Kuster, G. T., and Toksöz, M. N., 1974. Velocity and attenuation of seismic waves in two-phase media: Part I. Theoretical formulations. *Geophysics,* 39(5): 587-606.

Lachance, D. P., and Andersen, M. A., 1983. Comparison of uniaxial strain and hydrostatic stress pore-volume compressibilities in the Nugget sandstone. (SPE paper 11971) Society of Petroleum Engineers, Dallas, 8 pp.

Leighton, W., 1970. *Ordinary Differential Equations, 3rd Ed.* Wadsworth Publishing Co., Belmont, Calif., 287 pp.

Ling, C.-B., 1948. On the stresses in a plate containing two circular holes. *Journal of Applied Physics,* 19(1): 77-82.

Lo, T.-W, Coyner, C. B., and Toksöz, M. N., 1986. Experimental determination of elastic anisotropy of Berea sandstone, Chicopee shale, and Chelmsford granite. *Geophysics,* 51(1): 164-171.

Mackenzie, J. K., 1950. The elastic constants of a solid containing spherical holes. *Proceedings of the Physical Society of London,* 63(B1): 2-11.

Mann, R. L., and Fatt, I., 1960. Effect of pore fluids on the elastic properties of sandstone. *Geophysics,* 25(2): 433-444.

Marek, B. F., 1971. Predicting pore compressibility of reservoir rock. *SPE Journal,* 11(3): 340-341.

Matthews, C. S., and Russell, D. G., 1967. *Pressure Buildup and Flow Tests in Wells.* (SPE Monograph Volume 1) Society of Petroleum Engineers, Dallas, 172 pp.

Mavko, G. M., 1980. Velocity and attenuation in partially molten rocks. *Journal of Geophysical Research,* 85(B10): 5173-5189.

Mavko, G., and Nur, A., 1975. Melt squirt in the asthenosphere. *Journal of Geophysical Research,* 80(11): 1444-1448.

Mavko, G., and Nur, A., 1978. The effect of nonelliptical cracks on the compressibility of rocks. *Journal of Geophysical Research,* 83(9): 4459-4468.

Mesri, G., Adachi, K., and Ullrich, C. R., 1976. Pore-pressure response in rock to undrained change in all-round stress. *Géotechnique,* 26(2): 317-330.

Morlier, P., 1971. Description de l'etát de fissuration d'une roche à partir d'essais non-destructifs simples (Description of the state of rock fracturization through simple non-destructive tests). *Rock Mechanics,* 3(3): 125-138.

Murnaghan, F. D., 1951. *Finite Deformation of an Elastic Solid.* John Wiley & Sons, New York, 140 pp.

Murphy, W. F., 1984. Acoustic measurements of partial gas saturation in tight sandstones. *Journal of Geophysical Research,* 89(12): 11549-11559.

Muskhelishvili, N. I., 1953. *Some Basic Problems of the Mathematical Theory of Elasticity, 4th Ed.* Noordhoff, Groningen, 704 pp.

Newman, G. H., 1973. Pore volume compressibility of consolidated, friable, and unconsolidated reservoir rocks under hydrostatic loading. *SPE Journal of Petroleum Technology,* 25(2): 129-134.

Nishizawa, O., 1982. Seismic velocity anisotropy in a medium containing oriented cracks - transversely isotropic case. *Journal of Physics of the Earth,* 30(4): 331-347.

Norris, A. N., 1985. A differential scheme for the effective moduli of composites. *Mechanics of Materials,* 4(1): 1-16.

Norris, A. N., 1989. Stonely-wave attenuation and dispersion in permeable formations. *Geophysics,* 54(3): 330-341.

Nunziato, J. W., and Cowin, S. C., 1979. A nonlinear theory of elastic materials with voids. *Archive for Rational Mechanics and Analysis,* 72(2): 175-201.

Nur, A., and Byerlee, J. D., 1971. An exact effective stress law for elastic deformation of rock with fluids. *Journal of Geophysical Research,* 76(26): 6414-6419.

O'Connell, R. J., and Budiansky, B., 1974. Seismic velocities in dry and saturated cracked solids. *Journal of Geophysical Research,* 79(35): 5412-5426.

O'Connell, R. J., and Budiansky, B., 1976. Reply to Bruner's comment. *Journal of Geophysical Research,* 81(14): 2577-2578.

O'Donnell, T. P., and Steif, P. S., 1989. Elastic-plastic compaction of a two-dimensional assemblage of particles. *ASME Journal of Engineering Materials and Technology,* 111(4): 404-408.

Ostensen, R. W., 1983. Microcrack permeability in tight gas sandstones. *SFE Journal,* 23(6): 919-927.

Palciauskas, V. V., and Domenico, P. A., 1989. Fluid pressures in deforming porous rocks. *Water Resources Research,* 25(2): 203-213.

Peltier, B., and Atkinson, C., 1987. Dynamic pore pressure ahead of the bit. *SPE Drilling Engineering,* 2(4): 351-358.

Pettijohn, F. J., 1957. *Sedimentary Rocks, 2nd Ed.* Harper & Row, New York, 718 pp.

Pittmann, E. D., 1984. The pore geometries of reservoir rocks. In: D. Johnson and P. N. Sen (Editors), *Physics and Chemistry of Porous Media.* American Institute of Physics, New York, pp. 1-19.

Plona, T. J., 1980. Observation of a second bulk compressional wave in a fluid-saturated porous solid at ultrasonic frequencies. *Applied Physics Letters,* 36(4): 259-261.

Pollard, D. D., 1973. Equations for stress and displacement fields around pressurized elliptical holes in elastic solids. *Mathematical Geology,* 5(1): 11-25.

Raghavan, R. and Miller, F. G., 1975. Mathematical analysis of sand compaction. In: Chilingarian, G. V., and Wolf, K. H. (Editors), *Compaction of Course-Grained Sediments, Vol. I.* Elsevier, Amsterdam, pp. 403-524.

Rice, J. R., and Cleary, M. P., 1976. Some basic stress diffusion solutions for fluid-saturated elastic media with compressible constituents. *Reviews of Geophysics and Space Physics,* 14(2): 227-241.

Robin, P.-Y. F., 1973. Note on effective pressure. *Journal of Geophysical Research,* 78(14): 2434-2437.

Ruzyla, K., 1986. Characterization of pore space by quantitative image analysis. *SPE Formation Evaluation,* 1(3): 389-398.

Sack, R. A., 1946. Extension of Griffith's theory of rupture to three dimensions. *Proceedings of the Physical Society of London,* 58(6): 729-736.

Sadowsky, M. A., and Sternberg, E., 1947. Stress concentration around an ellipsoidal cavity in an infinite body under arbitrary plane stress perpendicular to the axis of revolution of cavity. *ASME Journal of Applied Mechanics,* 14(3): 191-201.

Sadowsky, M. A., and Sternberg, E., 1949. Stress concentration around a triaxial ellipsoidal cavity. *ASME Journal of Applied Mechanics,* 16(2): 149-157.

Saito, T., and Abe, M., 1984. A study on the distribution of pore shapes in crystalline limestone using the theory of composite materials. *Butsuri-Tanko,* 37(1): 15-26.

Salganik, R. L., 1973. Mechanics of bodies with many cracks. *Mechanics of Solids,* 8(4): 135-143.

Sampath, K., 1982. A new method to measure pore volume compressibility of sandstones. *SPE Journal of Petroleum Technology,* 34(6): 1360-1362.

Santarelli, F. J., Brown, E. T., and Maury, V., 1986. Analysis of borehole stresses using pressure-dependent, linear elasticity. *International Journal of Rock Mechanics and Mining Sciences,* 23(6): 445-449.

Savin, G. N., 1961. *Stress Concentration Around Holes.* Pergamon, New York, 430 pp.

Sawabini, C. T., Chilingar, G. V., and Allen, D. R., 1971. Design and operation of a triaxial, high-temperature, high-pressure compaction apparatus. *Journal Sedimentary Petrology,* 41(3): 871-881.

Scheidegger, A. E., 1974. *The Physics of Flow through Porous Media, 3rd Ed.* University of Toronto Press, Toronto, 353 pp.

Schmoker, J. W., and Gautier, D. L., 1989. Compaction of basin sediments: Modeling based on time-temperature history. *Journal of Geophysical Research,* 94(B6): 7379-7386.

Schreiber, E., Anderson, O. L., and Soga, N., 1973. *Elastic Constants and their Measurement.* McGraw-Hill, New York, 196 pp.

Seeburger, D. A., and Nur, A., 1984. A pore space model for rock permeability and bulk modulus. *Journal of Geophysical Research,* 89(B1): 527-536.

Shankland, T. J., and Halleck, P. M., 1981. Physical effects in rock under negative effective pressures: sound speeds and hydraulic diffusivity. In: *Rock Mechanics from Research to Application: Proceedings of the Twenty-Second U.S. Symposium on Rock Mechanics.* MIT Press, Cambridge, Mass., pp. 111-115.

Siegfried, R., and Simmons, G., 1978. Characterization of oriented cracks with differential strain analysis. *Journal of Geophysical Research,* 83(B3): 1269-1278.

Simmons, G., and Wang, H. F., 1971. *Single Crystal Elastic Constants and Calculated Aggregate Properties: A Handbook.* MIT Press, Cambridge, Mass., 370 pp.

Sneddon, I. N., 1946. The distribution of stress in the neighbourhood of a crack in an elastic solid. *Proceedings of the Royal Society of London,* A187: 229-260.

Sneddon, I. N., and Elliott, H. A., 1946. The opening of a Griffith crack under internal pressure. *Quarterly of Applied Mathematics,* 4(3): 262-267.

Sokolnikoff, I. S., 1956. *Mathematical Theory of Elasticity, 2nd Ed.* McGraw-Hill, New York, 476 pp.

Somerton, W. H., 1982. Porous rock-fluid systems at elevated temperatures and pressures. In: *Recent Trends in Hydrogeology.* (GSA Special Paper 189) Geological Society of America, Boulder, Colo., pp. 183-197.

Sprunt, E. S., and Nur, A., 1977. Destruction of porosity through pressure solution. *Geophysics,* 42(4): 726-741.

Sternberg, E., and Sadowsky, M.A., 1952. On the axisymmetric problem of the theory of elasticity for an infinite region containing two spherical cavities. *ASME Journal of Applied Mechanics,* 19(1): 19-27.

Swanson, R. K., Bernard, W. J., and Osoba, J. S., 1986. A summary of the geothermal and methane production potential of U.S. Gulf Coast geopressured zones from well test data. *SPE Journal of Petroleum Technology,* 39(13): 1365-1370.

Teeuw, D., 1971. Prediction of formation compaction from laboratory compressibility data. *SPE Journal,* 11(3): 263-271.

Terzaghi, K., 1936. The shearing resistance of saturated soils and the angle between the planes of shear. In: *Proceedings of the International Conference on Soil Mechanics and Foundation Engineering, Vol. 1.* Harvard University Press, Cambridge, Mass., pp. 54-56.

Thomsen, L., 1985. Biot-consistent elastic moduli of porous rocks: Low-frequency limit. *Geophysics,* 50(12): 2797-2807.

Toksöz, M. N., Cheng, C. H., and Timur, A., 1976. Velocities of seismic waves in porous rocks. *Geophysics,* 41(4): 621-645.

Tosaya, C., and Nur, A., 1982. Effect of diagenesis and clays on compressional velocities in rocks. *Geophysical Research Letters,* 9(1): 5-8.

van der Kamp, G., and Gale, J. E., 1983. Theory of earth tide and barometric effects in porous formations with compressible grains. *Water Resources Research,* 19(2): 538-544.

van der Knaap, W., 1959. Nonlinear behavior of elastic porous media. *Petroleum Transactions of the AIME,* 216: 179-187.

Venturini, W. S., 1983. *Boundary Element Methods in Geomechanics.* (Lecture Notes in Engineering, Vol. 4) Springer-Verlag, Berlin, 246 pp.

Walder, J., and Nur, A., 1986. Permeability measurements by the pulse-decay method: Effects of poroelastic phenomena and non-linear pore pressure diffusion. *International Journal of Rock Mechanics and Mining Sciences,* 23(3): 225-232.

Walsh, J. B., 1965a. The effect of cracks on the compressibility rocks. *Journal of Geophysical Research,* 70(2): 381-389.

Walsh, J. B., 1965b. The effect of cracks on the uniaxial elastic compression of rock. *Journal of Geophysical Research,* 70(2): 399-411.

Walsh, J. B., 1965c. The effect of cracks in rocks on Poisson's ratio. *Journal of Geophysical Research,* 70(20): 5249-5257.

Walsh, J. B., 1980. Static deformation of rock. *Journal of the Engineering Mechanics Division of the ASCE,* 106(EM5): 1005-1019.

Walsh, J. B., and Decker, E. R., 1966. Effect of pressure and saturating fluid on thermal conductivity of compact rock. *Journal of Geophysical Research,* 71(12): 3053-61.

Walsh, J. B., and Grosenbaugh, M. A., 1979. A new model for analyzing the effect of fractures on compressibility. *Journal of Geophysical Research,* 84(7): 3532-3536.

Walsh, J. B., Brace, W. F., and England, A. W., 1965. Effect of porosity on compressibility of glass. *Journal of the American Ceramic Society,* 48(12): 605-608.

Wang, C. Y., Mao, N. H., and Wu, F. T., 1980. Mechanical properties of clays at high pressure. *Journal of Geophysical Research,* 85(B3): 1462-1468.

Warren, N., 1969. Elastic constants versus porosity for a highly porous ceramic, perlite. *Journal of Geophysical Research,* 74(2): 713-719.

Warren, N., 1973. Theoretical calculation of the compressibility of porous media. *Journal of Geophysical Research,* 78(2): 352-362.

Warren, T. M., and Smith, M. B., 1985. Bottomhole stress factors affecting drilling rate at depth. *SPE Journal of Petroleum Technology,* 37(9): 1523-1533.

Watt, J. P., Davies, G. F., and O'Connell, R. J., 1976. The elastic properties of composite materials. *Reviews of Geophysics and Space Physics,* 14(4): 541-563.

Weinbrandt, R. M., and Fatt, I., 1969. Scanning electron microscope study of the pore structure of sandstone. In: *Rock Mechanics - Theory and Practice: Proceedings of the Eleventh Symposium on Rock Mechanics*. Society of Mining Engineers, New York, pp. 629-641.

White, F. M., 1974. *Viscous Fluid Flow*. McGraw-Hill, New York, 724 pp.

White, J. E., 1983. *Underground Sound: Application of Seismic Waves*. (Methods in Geochemistry and Geophysics, Vol. 18) Elsevier, New York, 284 pp.

Wilhelmi, B., and Somerton, W. H., 1967. Simultaneous measurement of pore and elastic properties of rocks under triaxial stress conditions. *SPE Journal*, 7(3): 283-294.

Wilkins, R. H., Simmons, G., Wissler, T. M., and Caruso, L., 1986. The physical properties of a set of sandstones, III: The effect of fine-grained pore-filling material on compressional velocity. *International Journal of Rock Mechanics and Mining Sciences*, 23(4): 313-325.

Willis, J. R., and Bullough, R., 1969. The interaction of finite gas bubbles in a solid. *Journal of Nuclear Materials*, 32(1): 76-87.

Wissa, A. E. Z., 1969. Pore pressure measurement in saturated stiff soils. *Journal of the Soil Mechanics and Foundations Division of the ASCE*, 95(SM4): 1063-1073.

Wyble, D. O., 1958. Effect of applied pressure on the conductivity, porosity, and permeability of sandstones. *Petroleum Transactions of the AIME*, 213: 430-432.

Yale, D. P., 1984. *Network Modeling of Flow, Storage and Deformation in Porous Rocks*. Ph.D. dissertation, Stanford University, 167 pp.

Zhang, J., Wong, T.-F., and Davis, D. M., 1990. Micromechanics of pressure-induced grain crushing in porous rocks. *Journal of Geophysical Research*, 95(B1): 341-352.

Zimmerman, R. W., 1984a. *The Effect of Pore Structure on the Pore and Bulk Compressibilities of Consolidated Sandstones*. Ph.D. dissertation, University of California at Berkeley, 116 pp.

Zimmerman, R. W., 1984b. The elastic moduli of a solid with spherical pores: New self-consistent method. *International Journal of Rock Mechanics and Mining Sciences*, 21(6): 339-343.

Zimmerman, R. W., 1985a. Compressibility of an isolated spheroidal cavity in an isotropic elastic medium. *ASME Journal of Applied Mechanics*, 52(3): 606-608.

Zimmerman, R. W., 1985b. Discussion of "The constitutive theory for fluid-filled porous materials," by N. Katsube. *ASME Journal of Applied Mechanics*, 52(4): 983.

Zimmerman, R. W., 1985c. The effect of microcracks on the elastic moduli of brittle solids. *Journal of Materials Science Letters*, 4(12): 1457-1460.

Zimmerman, R. W., 1986. Compressibility of two-dimensional cavities of various shapes. *ASME Journal of Applied Mechanics*, 53(3): 500-504.

Zimmerman, R. W., 1988. Stress singularity around two nearby holes. *Mechanics Research Communications*, 15(2): 87-90.

Zimmerman, R. W., 1989. Thermal conductivity of fluid-saturated rocks. *Journal of Petroleum Science and Engineering,* 3(3): 219-227.

Zimmerman, R. W., and King, M. S., 1985. Propagation of acoustic waves through cracked rock. In: *Research and Engineering Applications in Rock Masses: Proceedings of the Twenty-Sixth U.S. Symposium on Rock Mechanics.* Balkema, Rotterdam, pp. 739-745.

Zimmerman, R. W., and King, M. S., 1986. The effect of the extent of freezing on seismic velocities in unconsolidated permafrost. *Geophysics,* 51(6): 1285-1290.

Zimmerman, R. W., Haraden, J. H., and Somerton, W. H., 1985. The effect of pore pressure and confining pressure on pore and bulk volume compressibilities of consolidated sandstones. In: *Measurement of Rock Properties at Elevated Pressures and Temperatures,* (ASTM Special Technical Publication 869) American Society for Testing and Materials, Philadelphia, pp. 24-36.

Zimmerman, R. W., Somerton, W. H., and King, M. S., 1986a. Compressibility of porous rocks. *Journal of Geophysical Research,* 91(B12): 12765-12777.

Zimmerman, R. W., King, M. S., and Monteiro, P. J. M., 1986b. The elastic moduli of mortar as a porous-granular material. *Cement and Concrete Research,* 16(2): 239-245.

Abbreviations used

AAPG	American Association of Petroleum Geologists
AIME	American Institute of Mining Engineers
ASCE	American Society of Civil Engineers
ASME	American Society of Mechanical Engineers
ASTM	American Society for Testing and Materials
GSA	Geological Society of America
SPE	Society of Petroleum Engineers

Author Index

Subject Index